MENACING SKIES

MENACING SKIES

TEXAS WEATHER AND STORIES OF SURVIVAL

DAN HENRY

Cumulonimbus
PUBLISHING

MENACING SKIES
Texas Weather and Stories of Survival

ISBN   978-1-5445-0769-9   Hardcover
       978-1-5445-0767-5   Paperback
       978-1-5445-0768-2   Ebook

THREE-DIMENSIONAL READING
EXPERIENCE

Throughout this book you will find QR (Quick Response) codes that will open videos and animations to provide an even more engaging reading experience. The codes will automatically scan with both Android phones running on Android 8 and higher and iPhones (ios 11 or higher). The QR code below will take you to my website, where you will find links to additional weather resources and can leave comments about my book. Go ahead and grab your phone and follow the instructions below to make sure your device is compatible.

Scanning a QR Code

Simply launch the camera app and hold your device steady for 2–3 seconds toward the QR code. Once the QR code is read by your phone, a notification

banner will appear on the screen. Tap the notification to trigger the code's action. If you have an older phone that does not automatically scan the codes, there are several free QR code reader apps you can download at the App Store and Google Play store.

In the event that one of these video links is no longer accessible, you may be directed to an alternate link. If you encounter problems accessing the animation or video clip associated with a particular QR code, please visit my website and send me a message.

CONTENTS

INTRODUCTION

It was President's Day weekend in 1979, and my family was spending the weekend with my grandparents at their home in Reading, Pennsylvania. We had recently moved to the East Coast from sunny California and were about to experience our first real taste of winter weather. Just before going to bed, I flipped through all three of the late-night news channels to catch the latest forecast. The Philadelphia area meteorologists were all predicting 1–3" of snow by morning. My grandfather, sensing our excitement, had already retrieved the old Flexible Flyer sled from the basement and it stood by the door, ready to go. My sister and I planned to wake up early and hit the hill perched high above our grandparents' backyard for a few runs before we had to head home.

The next morning, my dad came into the bedroom to wake us up. The sly look on his face told me he knew something he wasn't telling us. He stepped into the room and told us to look out the window. I jumped out of bed, peeled back the curtains, and couldn't believe my eyes. Let's just say the forecast was a touch off. The yard was not blanketed by 2" of snow, or even three—it was buried under 2' of fresh powder, and the snow was still coming down. Ironically, the snowfall bounty had derailed our sledding plans. We ditched the sled for shovels and began the process of digging the car out of the driveway. Our next worry was getting home. This would undoubtedly be a long, grueling drive back to Delaware.

As it turned out, we wouldn't make it back home for a day and a half. Ten-foot-high snowdrifts shut down Highway 13, forcing us to spend the night in a hotel in Dover. The President's Day Blizzard did offer my sister and me an unexpected reward. Actually, two for me. We both got an entire week off from school, and I got instantly hooked on the weather.

After thirty years as a broadcast meteorologist, I am still in awe of the sheer power and unpredictability of weather. But I am equally amazed by the will of people to survive and endure the hardships faced after the winds cease and the floodwaters recede. Having covered Texas weather for more than seventeen years, I can tell you that the Lone Star State is not immune from any of the forces of nature. We've seen it all: hurricanes, tornadoes, flash flooding, drought, heat waves, arctic blasts, wildfires, windstorms, mega hailstorms, ice, sleet, and snow. Sometimes, we can witness many of these extremes within the span of twenty-four hours. Texas is also home to some of the fastest-growing metropolitan areas in the country, including the Dallas-Fort Worth Metroplex, the most populated urban center in "Tornado Alley." Development and sprawl have made North Texas a very large target, rendering us highly vulnerable to catastrophic weather events that threaten lives and property. The violent tornadoes that hit North Texas on December 26, 2015, stand as a grim reminder that dangerous weather can strike at any time of the year. Despite days of advance warning, many people were caught off guard by the deadly, out-of-season outbreak.

Truth be told, even with dual polarization radar at our fingertips and high-resolution computer models that simulate the atmosphere like never before, tornadoes and the storms that spawn them still have the power to perplex meteorologists. We don't know why some supercells can spin for hours and never produce a tornado, yet a seemingly innocuous storm can suddenly intensify and give birth to a long-track EF-4 twister. Our understanding of the atmosphere enables us to identify days that are ripe for giant hail, destructive winds, and torrential downpours that trigger flash flooding, but pinpointing precisely when and where these storms will strike remains well beyond the scope of our abilities. Record-setting extreme weather events and the inevitable debate on climate change that follows seem to command our attention more than ever before. Along with that heated discussion typically comes a great deal of hype and attention-grabbing headlines, yet little effort to dig deeper to unearth the facts and clear up the confusion.

I wrote this book for the casual weather observer and the die-hard weather junkie alike. After all, weather affects everyone. Some days, it presents a mere inconvenience—being caught in the rain without an umbrella or stuck in the bleachers for Friday Night Lights without a sweatshirt. But there are more than a handful of other days, especially in Texas, when the atmosphere goes haywire. On those days, a wrathful storm can have profound and lasting impacts. I have interwoven, with the science behind the weather, stories from real people like you and me who never expected they would lie squarely in the path of an EF-4 tornado, get swept up in raging floodwaters in the middle of the night, or be on board a jet taken down by a powerful microburst. In every disaster, there are stories of painful loss and mourning, heroism and bravery, and survival against all odds. Often, it is strangers who risk their lives as first responders and generously donate their time and resources to help fellow Texans in need.

I would like my readers to walk away with a better understanding of the driving forces behind our notoriously fickle weather, learn how to be better prepared for its sudden changes, and discover more about the impacts global climate change may have on our local weather. Lastly, my hope is that by reading these stories of survival, you'll be encouraged to know that the human spirit is still alive and well.

CHAPTER 1

DAY-AFTER CHRISTMAS TORNADO TRAGEDY

DECEMBER 26, 2015

)(

The only thing more terrifying than a large tornado hitting a highly populated area is one that strikes at night. Under the cloak of darkness, they are often not seen or heard until they are bearing down on you. When you think of Christmas, many images may come to mind, but the last thing would be witnessing a swarm of tornadoes decimating neighborhoods and turning lives upside down. On December 26, 2015, however, that unthinkable scenario played out across North Texas as powerful storms tapped into a volatile air mass, fueling a deadly outbreak that spun up twelve tornadoes over the span of eight hours, casting a long, dark shadow over what is supposed to be one of the most joyous times of the year.

Gary Tucker and his girlfriend, Amy Clark, were relaxing at home the day after Christmas when they got the itch to get out and do something. Eager to hit the stores for a little post-holiday shopping, they drove to Bass Pro Shop in Garland. It was a balmy afternoon and the sky was filled with dark, rolling clouds. Unusual, yes, but being a long-time Texan, Gary wasn't fazed. Gary and Amy spent about an hour inside the store before making their way to the registers. While they were checking out, Gary recalls looking out the exit doors and seeing torrential rain being blown sideways by very strong winds. By the time they had paid, the heavy downpours had ceased, and the wind had died down. As they headed to the exit doors, another shopper's phone received an alert for a tornado warning. It included a lengthy list of towns and cities, all of which Gary recognized, including Garland and neighboring Rowlett. Figuring the storm had subsided and considering he had heard about a thousand of these warnings before, he and Amy made their way out into the parking lot. It was 6:30 p.m. The sun had set an hour earlier and only a hint of twilight remained. The winds were now calm, but the eerie green skies were still cause for concern. As they got into the truck, Gary asked his girlfriend if she was hearing a sound off in the distance. He made a slow, deliberate loop around the parking lot. Listening intently with the windows rolled down, he turned to Amy and said, "Now do you hear it?" This time, she did. It was still faint and distant, but it sounded like a train.

Gary naturally assumed that if something out of the ordinary was going on, it would be all over the radio. He tuned in to a local station to get a weather update. The station was not broadcasting any alerts and there were no reports of a confirmed tornado or any damage. Contemplating what to do, he recalled that the warning he had heard as they left the store mentioned a very large area. The weather was still eerily quiet. There was no rain and the winds were calm. He figured if there was a tornado, it had most likely passed them by, and he felt it was safe to start heading home.

They got on the Interstate 30 service road and headed west. Despite the seemingly tranquil weather, lightning was still casting streaks of light across the night sky. It wasn't long before they reached the ramp that would take them to the President George Bush Turnpike. As they came around the corner and approached the overpass, the sky exploded with a massive electrical flash of light. With the night sky suddenly illuminated before them, they saw the tornado. It was maybe a hundred yards ahead of them. The powerful winds were hurling huge pieces of wood in all directions. A section of guardrail went rocketing through the air. Amy urged Gary to back up, but it was too late. They could see the headlights of another truck coming right behind them. Within seconds, their vehicle was being blasted by debris. Gary put the truck in park and yelled, "Get down, get down!" Moments later, Gary said it felt like a dinosaur had put its foot on the back of his Toyota Tundra and was shaking it violently. Amy felt herself being lifted up out of her seat. In an instant, all the glass shattered and the airbags deployed. The last thing Gary recalls before blacking out was his face hitting the gearshift. The force of the impact caused him to bite through his tongue. All Gary remembers is that the deafening noise went completely silent. At that point, he thought he had died. He experienced a sense of tranquility overcome him and the intense pain he had been feeling just seconds earlier vanished. The chaos all around him suddenly stopped and it felt as if he was being lifted into heaven.

View from Rockwall Harbor *(KDFW-FOX 4 News)*

When he came to, the truck was resting on the driver side about 30' off the service road. Bloodied and still dazed, Gary collected himself and checked on Amy. Her arm was bleeding, but she was otherwise okay. Unbuckling himself, Gary carefully climbed over the seat and made his way out the back window, while Amy was assisted out of her window by an off-duty fireman and the driver who had been right behind them. Once out of the badly beaten truck and out of danger, they immediately sensed just how lucky they were to be alive. The scene was total pandemonium. It looked

like a war zone. A long section of guardrail was now sticking straight up after having pierced the ground, and a battered riding lawn mower was lying in the median. By far, the most sobering sight were the mangled remains of vehicles that had been tossed like toys. Only yards from where his truck came to rest

i-30/PGBT Garland *(KDFW-FOX 4 News)*

after going airborne, landing, and then rolling several times, he could see several crushed cars and trucks that had been thrown from high above on the I-30 overpass and landed below. They would later learn that nine people lost their lives there. Two were found inside their vehicles, while another seven victims were ejected and spread out over hundreds of yards.

Minutes later, a police officer checked on Gary and Amy. He told them an apartment complex just across the highway had been leveled and that they were probably not going to get an ambulance for quite a while. He instructed them to take the back way to the hospital if they needed medical help. A couple of Good Samaritans gave them a ride home and, from there, Gary's brother took them to the emergency room. Doctors informed Gary that he had sustained a broken jaw and Amy suffered a deep gash in her arm that required several stitches. Both also had several lacerations to the face from flying glass. After what they had just witnessed, they were extremely grateful to be alive.

Not far away, and just across Lake Ray Hubbard, Pam Russell and her wife, Dacia, had just finished dinner. They were aware of the threat of severe weather but figured, at most, they would get some hail and high winds. They had been monitoring live TV coverage and taking occasional peeks outside but were under the impression that the storm moving north from Sunnyvale would track on the west side of Lake Ray Hubbard. What they didn't know was that a deadly tornado had just slammed head-on into unsuspecting travelers on the highway and was now hurtling northeast, putting them squarely in its path.

The couple had just moved into their house in Rowlett the day after Thanksgiving. Drawn by the mature live oaks in the front yard and the proximity to beautiful Lake Ray Hubbard, it felt like the perfect place to call home. They had invited some friends over and were proudly sharing their plans for remodeling the interior of the house and sprucing up the landscaping. After dinner, Pam began cleaning the dishes while Dacia and their friends went outside to scan the sky once again. Looking to the south, they saw bright flashes on the horizon, but this was not lightning. Seconds later, an exploding transformer lit up the sky and left no doubt that what they were seeing was an approaching tornado. The transformer it had just wiped out was located close to a kayak business on the lake—a mere half a mile away. The immense tornado was approaching quickly, and they knew they only had precious seconds to take shelter.

Charging into the house, they yelled, "Take cover!" Pam grabbed a throw off the sofa, called for her cats, and they all raced into the guest bathroom, which was located in the center of the house. Pam recalled that it felt like the second they shut the door, the tornado hit, and they immediately felt the "power" of Mother Nature. Dacia wedged her feet against the door in an attempt to keep it shut, but the wind was too powerful. It was whistling through the house and pushing open the bathroom door. They could hear the banging, crackling, and crumbling all around them. The noise was deafening. Pieces of insulation swirled all around them and hit them in the face. Pam recalls thinking, "This is it. We are not going to walk away from this." At that moment, she comforted herself with the words, "If this was God's plan, it's God's plan."

Rowlett Destruction *(Volkan Yuksel)*

The tornado's fury came down on their home and their bodies so intensely that enduring it even for a minute seemed like an eternity. When the wind finally died down, it was pitch black and everything went dead silent. Remarkably, none of them had sustained any injuries. They looked up and saw the sky. The roof was gone. As they got their bearings, they immediately smelled gas. They knew they had to get out of the house, but there was debris everywhere and they could not see anything. The walls had all collapsed and one of the larger live oaks was now in what was left of their living room. Within minutes, they heard the voice of a neighbor calling from the street. He was asking if they were alright and if they needed assistance. In the furious scramble to take cover, Pam had forgotten to put a pair of shoes on. She asked him if he could get her a pair. Within minutes, he returned with some shoes and, very slowly and carefully, they all made their way out.

The devastation they saw in their neighborhood was indescribable. The place that they called home for a mere thirty days was leveled. They began trying to call family to let them know they were alive, but cell phone service was very sketchy. They finally were able to reach Pam's brother. Extremely relieved to hear their voices, he told them he was coming to get them. He got in his truck and drove as far as he could before downed trees and

Rowlett Home in Ruins *(Hollie Hernandez)*

power lines blocked his way. Calling out his name as they walked through a maze of debris that was scattered everywhere, they somehow managed to find each other. After some long embraces, they all piled into the truck and made their way back to Pam's brother's home. The next day, they returned to their devastated neighborhood to discover some good news amidst the tragedy; they found their beloved cats hiding in what was left of the air conditioning ducts. Their home was in ruins, both vehicles were totaled, but what mattered more than anything was that they had all survived.

ATMOSPHERE WAS RIPE FOR AN OUT-OF-SEASON OUTBREAK

The FOX 4 Weather Team provided live coverage of the December 26, 2015, tornado outbreak from beginning to end. As meteorologists, we saw the ingredients for a severe weather outbreak a couple of days prior to the event but still were skeptical given the time of year. While not unprecedented, stormy weather of this magnitude is very unusual in the winter season. The

Storm Prediction Center (spc) had issued a "slight risk" of severe weather the night before, which portends isolated storms packing wind gusts of sixty miles an hour or greater and hail the size of quarters or larger. We knew the air was unusually moist because the Gulf of Mexico water temperatures were still pretty warm. There had not been any big pushes of cold air yet that winter. In fact, with an average temperature of 53.7°F, December 2015 was the second warmest on record for the Dallas-Fort Worth area. On the morning of the December 26, it became apparent that the moisture content was even greater than originally forecast, with dew point readings near 70°F. These sultry conditions are common in spring in Texas but certainly not in late December. Additionally, the upper-air disturbance coming in from the west had grown stronger and would arrive during the late afternoon as a cold front approached from the west.

The spc recognized these factors and raised our risk level to "enhanced" at seven o'clock that morning. On "enhanced risk" days, you expect greater severe storm coverage with varying levels of intensity. Our concern had to do with whether people in North Texas would take this seriously. Even storm-savvy North Texans were not accustomed to this type of heightened severe weather alert this time of year and delivering the message during the holidays would be more difficult than usual with fewer people tuned to local television.

However, this day was quickly turning. Clouds were parting early, and the sun broke out along the I-35 corridor. The additional heat from the now-unobstructed sun sent temperatures soaring to 80°F. More evidence that this would be an active day was provided by the noontime weather balloon launch. Weather balloons can reach altitudes of over 100,000' above Earth, as they carry a lightweight box called a radiosonde. As the radiosonde rises, its weather sensors measure and transmit values of temperature, pressure, and relative humidity. This data is displayed in what is called an atmospheric sounding. Equipped with GPS, weather balloons can also provide information on wind speed and direction by their flight path and rate of ascent. This special balloon launch revealed that instability in the atmosphere was three times the normal value, and with the added heat and moisture, the cap was nearly gone. The cap, also called a "lid," is a layer of warm, stable air, usually around 4,000' to 6,000' above the ground that will often prevent air from rising above it, inhibiting thunderstorm formation. The atmospheric sounding also showed that the low-level winds were very strong and turning clockwise with height. All of these red flags prompted the SPC to issue a tornado watch at 12:35 p.m. for all of North Texas.

The initial storms developed south of the Metroplex around one thirty in the afternoon. They formed quickly and raced north at speeds of forty to fifty miles an hour. Meteorologists Evan Andrews and Ali Turiano were already on the air, tracking these formidable storms. The radar indicated nearly every storm as a potential hail or tornado producer. From 1:43 p.m. until 4:52 p.m., the National Weather Service issued eighteen tornado warnings, all of which were based on radar-indicated rotation. While none of these stood out as particularly threatening, meteorologists cannot afford to take chances with storms that are moving and developing this quickly. The first cell that really commanded our attention was in Ellis County, just south of Dallas. A tornado warning was issued at 4:54 p.m. for northwest Ellis County. It stated that the storm was moving north at forty miles an hour and would be near Waxahachie at 5:20 p.m., then track near Midlothian, Glenn Heights, Red Oak, and Ovilla around 5:25 p.m.

FOX 4 storm chaser Gene Yates lives in Glenn Heights and was waiting on the storm at Longbranch Elementary on FM 1387 in Midlothian. "I was actually deleting old videos to make room on my camera and had my window rolled down when I started to hear the roar. About the time I looked up, there were power flashes in front of me. I immediately drove west to get out of the way as it wasn't clear as to where it was moving." The tornado passed

Glenn Heights Damage *(Gene Yates)*

behind him, crossing the intersection of Ovilla Road and FM 1387. He then followed it as it moved northeast toward Glenn Heights. "I turned north on Hampton. My gut dropped as I could see it was moving through my neighborhood and the tornado was quite strong, but I had to try and keep my com-

posure." All he could think about was getting home as quickly as possible to check on his dad, a Glenn Heights police officer. "I didn't know it at the time, but when the tornado was moving into the neighborhood my dad was literally running into the house. He told me [that] as soon as he got into the closet, he thought, 'This was it.'" After a ten-minute "white-knuckle" drive that had him taking multiple detours around debris-filled streets, Gene arrived to discover that his home and his father were both untouched. The tornado had missed their house by one block.

Reports from other eyewitnesses were telling us what Gene already knew. This was a strong tornado. As the storm moved out of Ellis County into Dallas County, debris was being indicated on radar. At 5:28 p.m., a tornado warning was issued for central Dallas County. The storm was now over Lancaster and was traveling north at forty miles an hour. This put much of the heart of Dallas County, including all of our staff at FOX 4 in downtown Dallas, in the path of a dangerous storm.

As the ominous storm was rapidly approaching the city of Dallas, coworkers started asking if we intended to continue our live coverage or take shelter. It certainly was a valid concern and something we needed to take into careful consideration. If we took cover in the basement, could we continue our coverage, talking over radar from the WAPP (the FOX 4 weather app)? Would our microphones even work in the basement? We took a minute to examine the radar and factor in the storm's trends and its trajectory to determine if we were likely to be hit. Fortunately for us, the rotation in this storm was weakening as it approached downtown. Gene confirmed that the tornado had lifted. Our immediate thoughts were those of extreme relief. We had just dodged a bullet. But that solace lasted for all of five minutes, as we watched a storm in southeast Dallas County start to explode.

The rotation on the new storm spun up quickly as it approached Balch Springs. At 6:39 p.m., the National Weather Service issued a tornado warning for northeastern Dallas County, western Rockwall County, and southern Collin County. The warning stated that, "The dangerous storm would be near Sunnyvale around 6:45 p.m., then Dallas, Garland, Rowlett, Sachse, and Richardson around seven in the evening, and into Murphy around 7:05 p.m." Just minutes after broadcasting the warning, reports of a huge tornado flooded the newsroom. These reports were confirmed moments later, with several viewer photos shared via social media showing a wedge tornado near Lake Ray Hubbard, its massive funnel illuminated by lightning. Video from North Texas Tollway Authority (NTTA) cameras showed the tornado from multiple angles as it approached and then crossed Interstate 30 in Garland. Scan the QR code to watch it.

Garland Tornado Video *(NTTA)*

The radar image was equally frightening. It displayed a doughnut-hole appearance as it was crossing I-30 into Rowlett, just east of Dallas. The doughnut appearance is actually the combination of precipitation and debris spinning around the hole that marks the center of the tornadic circulation. On the air, we were very frank. "Folks, this doesn't look good at all. This is one of the largest circulations we have ever seen on radar and it will be devastating."

By this time, I had joined Evan and Ali in the weather center. It was an all-hands-on-deck situation and the outbreak was far from over. We felt helpless watching the outbreak unfold, knowing everyone in its path would be impacted severely and there would likely be loss of life. In our position, we try our best to put those feelings aside, remain focused, and do our job in order to best inform those potentially at risk.

As the Garland/Rowlett supercell barreled north into Collin County, National Weather Service meteorologists issued a new tornado warning at 7:13 p.m. for eastern Collin County and northwestern Hunt County. It read, "Radar is still tracking a strong rotation near Wylie moving northeast at fifty miles an hour. This is an extremely dangerous storm and has caused extensive damage and possibly injuries near Rowlett." To complicate matters, this was not the only potentially deadly storm on radar. Tornado warnings continued for portions of Ellis, Navarro, and Kaufman counties, so we were pressed to ensure we devoted time on the air to cover each storm.

Some of the vital components to accurate and timely storm coverage are the instantaneous weather and damage reports we get from our National Weather Service chat sessions. Storm chasers, amateur radio operators, law enforcement, emergency managers, and numerous other city officials all share information in this chat room. This night we saw more activity than ever before. Around seven thirty in the evening, a report came in of people trapped underneath a collapsed building in Copeville. A few minutes later, we received news of a gas station in Nevada that had been leveled.

Upon receiving these reports, the National Weather Service reissued the warning with even stronger wording. "At 7:34 p.m., a confirmed large and extremely dangerous tornado was located eight miles north of Farmersville or seventeen miles east of McKinney moving northeast at thirty-five miles an hour. This is a particularly dangerous situation. You are in a life-threatening situation. Mobile homes will be destroyed. Considerable damage to homes, businesses, and vehicles is likely, and complete destruction is possible. The tornado will be near Celeste around 7:45 p.m., Leonard around 7:50 p.m., Bailey around eight in the evening, and Bonham and Dodd City around 8:15 p.m."

By now, North Texans were glued to their TV sets and acutely aware of the ongoing storms. After seeing photos of the tornado and its destructive power, people were compelled to take cover. The storm still maintained strong rotation and was racing north at fifty miles an hour. The problem with storms that move into Dallas's distant suburbs is that radar can no longer capture the lower portions of the storm. This means knowing whether or not a tornado is still occurring depends almost entirely on getting ground truth reports from trained spotters. The last tornado warning was issued at 8:44 p.m. for Fannin County. Fortunately, at that point, the storms were weakening and were more of a heavy rain threat than anything else.

DELIVERING THE SAD NEWS OF LIVES LOST

Our live coverage continued on *FOX 4 News at Nine O'Clock*. We were now receiving the first reports of confirmed fatalities. Having to share that terrible news on the air was very difficult. As meteorologists, we strive to deliver accurate and timely information to our viewers, especially in life-threatening situations like this one. By this time of night, our news crews were sharing live pictures of the devastation as well as the first accounts from storm survivors. Veteran reporter James Rose was one of the first on the scene. "I just heard a noise I'd never heard in my life," a woman told Rose. "My kids started screaming." The woman and her children survived the tornado in Rowlett, but the entire roof of their two-story home was ripped off.

As the sun rose on December 27, the morning light revealed the true scope of the disaster. National Weather Service survey crews were dispatched all across North Texas to examine the damage. A total of twelve tornadoes were confirmed. The strongest was the EF-4 tornado with winds estimated up to 180 miles an hour that tracked from near Sunnyvale north through Garland and Rowlett. It killed ten people. Two other tornadoes were rated EF-2 or higher, including an EF-3 with winds around 150 miles an hour that carved an eight-and-a-half-mile path of destruction through Midlothian, Ovilla, and Glenn Heights. Miraculously, it caused no fatalities. The other three fatalities were caused by two separate tornadoes in Collin County. An EF-2 tornado was responsible for two deaths in Copeville, and an EF-1 killed one person in Blue Ridge.

The National Weather Service issued a total of thirty-five tornado warnings on December 26 that undoubtedly helped save many lives. Nearly 450 homes in Rowlett alone were damaged or destroyed, yet there was only one fatality. It is sobering and disheartening to know that thirteen people did perish in the storms. The events of that night have had a permanent effect on those who lost loved ones unexpectedly, and even those who survived will be changed forever.

COPING WITH LIFE AFTER THE TORNADO

Pam Russell, who lost her Rowlett home to the tornado, offers perspective, "It gives you a tremendous appreciation for life and what's truly important, especially time with family and friends. Material things are replaceable. Even with treasured belongings, you still have memories. We are here [on earth] to build relationships with people." She believes people who go through disasters like this suffer from post-traumatic stress disorder (PTSD) and sometimes do not realize it until months later. She encourages storm survivors who may be struggling with PTSD to ask for help. "Don't be afraid to let people help you. Lean on friends and family. Be honest with your feelings." Surviving the tornado has also strengthened her faith. She believes that the tornado hitting the day after Christmas, rather than at some other time of year, may have been a bit of a blessing. A few families in her neighborhood were gone over the holiday break and came back to find their homes demolished…Pam believes that had those families been home when the tornado hit, they may not have survived.

Gary Tucker, who survived after driving into the tornado, realizes he was extremely lucky. "I have had open heart surgery. I have been in another major car accident, and I survived this tornado. Apparently, it's not my time."

Pam has traversed a much more difficult road to recovery. Rarely a day goes by that she doesn't think about the tornado, and she attended counseling for a year and a half to help her deal with the painful memories and the severe emotional distress.

Gary also experiences frequent flashbacks. "I feel very bad for the people who lost their lives. They drove right into the tornado. They had no idea what they were driving into." In hindsight, Gary knows he should have heeded the warning and not driven home. It is a mistake he vows to never repeat again.

The twelve confirmed tornadoes would bring the total to seventy-six for 2015 in North Texas, breaking the previous annual record of seventy-three tornadoes in 1994. The December 26 outbreak is proof positive that tornadoes are not just a seasonal phenomenon limited mostly to springtime. Severe weather can strike at any time of day and during any month of the year in North Texas. But the same can be said about the weather in general in North Texas. It is just as common to see an 80°F day in winter as it is to see freezing rain or snow flurries. Sometimes, you may see both within twenty-four hours.

CHAPTER 2

WAIT FOR IT

THE JEKYLL AND HYDE NATURE
OF NORTH TEXAS WEATHER

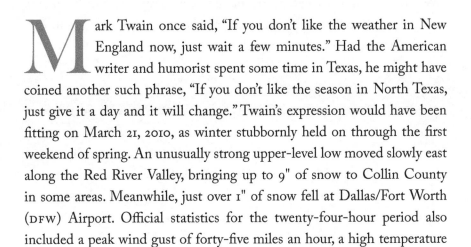

Mark Twain once said, "If you don't like the weather in New England now, just wait a few minutes." Had the American writer and humorist spent some time in Texas, he might have coined another such phrase, "If you don't like the season in North Texas, just give it a day and it will change." Twain's expression would have been fitting on March 21, 2010, as winter stubbornly held on through the first weekend of spring. An unusually strong upper-level low moved slowly east along the Red River Valley, bringing up to 9" of snow to Collin County in some areas. Meanwhile, just over 1" of snow fell at Dallas/Fort Worth (DFW) Airport. Official statistics for the twenty-four-hour period also included a peak wind gust of forty-five miles an hour, a high temperature of 62°F, patchy fog, partial sunshine, and 1.17" of rainfall.

While certain areas of the country may experience the variety of weather we do in Texas, very few see those extremes unfold so quickly. Relatively flat terrain and unobstructed views of the horizon allow you to see the weather coming for miles, and observe these changes happen before your eyes. Perhaps it is the view of the wide-open sky that makes many North Texans amateur weather observers by default. We witness breathtaking sunrises and unforgettable sunsets. We marvel as towering cumulonimbus clouds build miles into the sky and cast bolts of lightning in all directions. We often feel these changes taking place too. A balmy southerly breeze one minute can abruptly shift to a stinging northerly wind the next. Why does North Texas weather seem to always turn on a dime?

IT'S ALL ABOUT THE GEOGRAPHY

The single biggest factor is location. Situated in the Southern Plains with the majestic Rocky Mountains to our west, well over a thousand miles of wide-open prairies to the north, and the Gulf of Mexico to the south, we sit center stage as cold, dry air moves south frequently colliding with warm, moist air drawn northward from the Gulf. John Nielsen-Gammon, who has served as the Texas State Climatologist since 2000, says the Jekyll and

Hyde nature of North Texas weather is, "all about the geography." He further elaborates, "The Rockies make it very hard for Pacific air masses to move in from the West, which would be a moderating influence." Consider the city of Los Angeles, located at a similar latitude to Dallas and Fort Worth, but whose climate is stabilized due to its close proximity to the Pacific Ocean. Southern California winters are milder, and summers are cooler because they are dominated by air masses blowing in off the Pacific. This is why the average high in Los Angeles is 68°F in January, compared to 56°F in the Dallas-Fort Worth area. In August, the average high in LA is a comfortable 84°F compared to a sweltering 96°F in DFW—it's no wonder the Cowboys choose to spend a good portion of their training camp in Southern California come August.

North Texas is also subject to more frequent temperature extremes. Once again, the Rocky Mountains play a major factor. Peaking at over 14,000', this mountain range acts as a barrier to direct all of the cold, dry, continental polar (cP) and continental arctic (cA) air masses that originate in Alaska and northern Canada southward. The channeling effect of the mountains tends to enhance the severity of cold fronts in winter. Often called "Blue Northers," these cold fronts can move at breakneck speed, dropping temperatures by 30°F or more in just two hours and over 50°F in less than twenty-four hours. One of the biggest drops in recent years occurred on March 1–2, 2014. The high at DFW on March 1 was 81°F Then, an Arctic cold front barreled south of the Red River and temperatures plummeted, dropping over 60°F. The next morning, DFW recorded a low of 19°F.

Just as bitterly cold air encounters no geographical roadblocks on its long journey south from the Arctic Circle, southerly winds can freely transport moist maritime tropical (mT) air masses that originate over the Gulf of Mexico and shift northward. Warm fronts typically identify the leading edge of this warm, humid air. The Alps in Europe and the Himalayas in Asia are oriented east-west and keep these air masses separated. The Rockies stretch north-south, allowing the air masses to clash frequently, which says Nielsen-Gammon, "makes for highly changeable weather, especially in winter, because you are basically alternating between weather coming from the general direction of the Polar regions versus the Tropics."

EARTH'S TILT IS THE REASON FOR THE SEASONS

The sun's uneven heating of the earth's surface causes large differences in temperature around the globe. The earth's axis is tilted 23.5 degrees off vertical as it orbits the sun. In summer, when the Northern Hemisphere is tipped toward the sun, we experience stronger rays and longer days. In winter, when the Northern Hemisphere is tipped away from the sun, we experience weaker rays and shorter days. Think of the sun's rays as light from a flashlight. If you aim the flashlight directly at a wall (imitating the sun when it's directly overhead), the light is bright and concentrated. During summer in North Texas, the sun is about seventy-five to eighty degrees above the horizon, allowing for intense heating. Now, tilt the flashlight at an angle, to mimic the sun positioned low in the sky. The light spreads out and is much less concentrated. In a North Texas winter, the sun only rises about forty-five to fifty degrees above the horizon, so its rays are much weaker, and they heat less efficiently. The effect is amplified as you go from the equator toward the poles. The sun's rays are strong in tropical regions and heat the

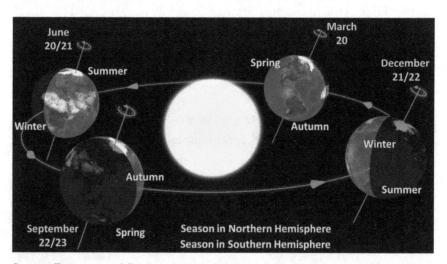

Seasons, Equinoxes, and Solstices *(Wikimedia Commons/Tau'olunga)*

air effectively. In polar regions, the opposite is true. This unequal heating around the globe creates differences in air pressure that stir up wind and ultimately drives our weather.

THE AIR ABOVE CREATES PRESSURE BELOW

Billions of air molecules are constantly pressing against everything they touch, including our bodies. We are not crushed by their cumulative pressure because billions of molecules inside our bodies are pushing outward with equal force. The air molecules above us are surprisingly heavy. A one-inch by one-inch column of air from the ground to the top of the atmosphere weighs 14.7 pounds. The weight of the air molecules acts as a force as it is pulled by gravity and presses down on the earth. That pressure is called atmospheric pressure, or air pressure, and the air exerts 14.7 pounds per square inch (psi) of pressure at the earth's surface.

The standard unit of pressure found on weather charts is the millibar (mb). Atmospheric pressure is commonly measured with a barometer, which contains a column of mercury in a glass tube that rises or falls as the weight of the atmosphere changes. Higher atmospheric pressure forces the mercury to rise, while lower atmospheric pressure causes the mercury to fall. At sea level, the standard value for atmospheric pressure is 29.92" of mercury (Hg) or 1,013.25 mb.

LAYERS OF THE ATMOSPHERE

The atmosphere is divided into five layers. It is thickest near the surface of the earth and thins out with height until it merges with space. The troposphere is the first layer above the surface and is where nearly all weather occurs. Most clouds are observed here, as 99% of the water vapor in the atmosphere is in the troposphere. Atmospheric pressure drops and temperatures get colder as you climb higher in the troposphere. Because there are fewer air molecules at higher altitudes, there is less pressure from the weight of the air above. At an altitude of 18,000' (about 4,000' higher than Pike's Peak), the air pressure is roughly 500 mb, or half of the sea-level pressure. At an elevation approaching that of the summit of Mount Everest (just over 29,000'), the air pressure is about 300 mb. A climber at the summit of the highest peak on Earth is above nearly 70% of all air molecules.

The tropopause is the boundary separating the troposphere from the stratosphere. The stratosphere is home to the "ozone layer," the large concentration of ozone that absorbs much of the sun's ultraviolet radiation and heats the atmosphere. This explains why the air temperature increases with height in this layer, producing a "temperature inversion." The inversion keeps the stratosphere stable and reduces turbulent air motions. Commercial jet aircraft fly in the lower stratosphere to avoid the turbulence that is common in the troposphere below. The stratopause marks the top of the stratosphere. At this level, the air pressure is only 1 mb. This means that 99.9% of the atmosphere is contained in the stratosphere and troposphere.

Above the stratosphere is the mysterious mesosphere. Extending from thirty-one to fifty miles above the earth's surface, the mesosphere is out of reach of weather balloons and aircraft, while satellites orbit above it. This prevents us from taking any measurements in this layer. The air at this altitude is extremely thin and the temperature decreases with height throughout this layer. In fact, the coldest temperatures in Earth's atmosphere, about −130°F, are found near the top of the mesosphere. Most meteors burn up in the mesosphere. This layer is nearly devoid of clouds due to the lack of moisture, however, wavy bluish-white "noctilucent clouds" sometimes form in the mesosphere near the North and South Poles.

Directly above the mesosphere is the thermosphere. Although the thermosphere is considered part of Earth's atmosphere, the air density at this altitude is so low that we normally think of most of the thermosphere as outer space. The Karman Line is the most widely (though not universally) accepted boundary that defines the altitude where the atmosphere ends and space begins. The space shuttle and the International Space Station both orbit Earth within the thermosphere.

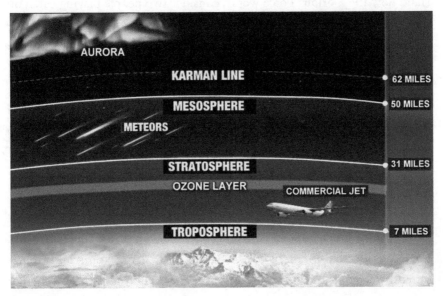

AURORA

KARMAN LINE .... ● 62 MILES

MESOSPHERE ● 50 MILES

METEORS

STRATOSPHERE ● 31 MILES

OZONE LAYER   COMMERCIAL JET

TROPOSPHERE ● 7 MILES

Adapted from NOAA/NESDIS image

Finally, the atmosphere merges into interplanetary space in the exosphere. Some scientists consider the thermosphere the uppermost portion of our atmosphere. Only the lightest gases (hydrogen, helium, carbon dioxide, and atomic oxygen) occur in the exosphere, barely held to the planet by gravity.

WIND: THE GREAT EQUALIZER

The differences in temperature from the poles to the tropics create pressure variances. Warm air rises because it is less dense. As it rises, it creates an area of low pressure at the surface. In contrast, the cold sinking air produces an area of high pressure. When differences in horizontal air pressure exist, there is a net force that causes the air to move to restore a pressure balance in the atmosphere. This force, called the "pressure gradient force," is directed from higher pressure to lower pressure and causes the wind to blow. Once air is in motion, it is deflected in its path by the "Coriolis Force."

The Coriolis effect describes an apparent force due to the rotation of the earth. In the Northern Hemisphere, this turns wind to the right. To understand how the Coriolis Force works, consider two people playing catch on opposite sides of a merry-go-round spinning counterclockwise (same direction earth spins as viewed from above the North Pole). If a ball is thrown straight across to the person on the other side, the ball will appear to veer to the right. A force really is not acting on the ball to make it go off course; it merely appears to curve because of the merry-go-round's movement. The same effect occurs on the rotating earth. The Coriolis Force is what causes wind to blow counterclockwise around low pressure and clockwise around high pressure in the Northern Hemisphere—the opposite holds true in the Southern Hemisphere.

Wind plays a crucial role in correcting the imbalances in temperature that occur over Earth's surface by redistributing heat energy. Warm air is transferred from tropical regions toward both poles, while cool air from higher latitudes is directed toward the Equator. Wind also transports water vapor from the oceans to inland areas where the vapor condenses, and allows greater precipitation over land surfaces. Without this moisture transport, inland areas would be drier.

GULF BREEZES TAP MOISTURE

North Texas is known for its strong, gusty winds. The mean annual wind speed is 10.5 miles an hour in the DFW area. That is actually a tad higher than a Midwestern city famed for its gales. Chicago has a mean annual wind speed that clocks in at 10.3 miles an hour with the aid of strong breezes generated by nearby Lake Michigan. Just as the weather in the Windy City is dependent on Lake Michigan breezes, our weather in North Texas is tied to the prevailing winds. The dominant surface air flow in North Texas is from the south. When winds blow from the south, they tap into moisture from the Gulf of Mexico. Since moisture fuels spring thunderstorms and winter storms, much of our annual rainfall is tied to these winds.

With the exception of the heart of summer, mid-July through the end of August, precipitation in North Texas is fairly evenly distributed throughout the year. However, rainfall can vary substantially from year to year. Since official record keeping began in 1899 for the DFW area, we have had eight years with rainfall totals less than 20" and seven years when rainfall exceeded 50". The driest year on record was 1921, with only 17.91" of rainfall. The wettest occurred in 2015, with a staggering 61". Some of these wild fluctuations can be tied to changes in global weather patterns triggered by El Niño and La Niña. We will explore these and other climate oscillations in later chapters.

Spring in North Texas typically delivers the greatest number of days with rainfall. This is no surprise when you consider the frequency of thunderstorms that produce heavy downpours. Averaging 4.90" of rainfall, May is the wettest month. The frequent, heavy rains, can wreak havoc for large outdoor events. The organizers of Mayfest in Fort Worth can certainly attest to that. The tail end of our spring storm season is characterized more by nocturnal storms. These storms flare up in the High Plains of southwest Kansas and the Panhandles of Texas and Oklahoma during the afternoon and early evening. Once these storms organize into a complex, they can persist for many hours, traveling 300 miles or more, as they move east and southeast.

It is usually well after midnight when these complexes roll south of the Red River and deliver an unpleasant, early wake-up call for North Texans.

As we head into late June and July, a ridge of high pressure builds overhead and dominates our weather. Beneath these high-pressure domes, the air sinks, compresses, and warms. A similar thing happens when you use a bicycle pump. Air is compressed as it is pumped into the tire and the outside cylinder gets warmer. In nature, this sinking air produces what is called a "subsidence inversion." This warm, dry layer of air acts like a lid and prevents air from rising above it, thus suppressing thunderstorm development. Occasionally, the high-pressure ridge will weaken, allowing for isolated thunderstorms to pop up during the afternoon. These storms tend to be brief and slow moving but can produce some localized heavy rain and occasionally strong winds as they weaken and collapse. By August, we usually experience wall-to-wall sunshine with little hope of any rain. August is typically the driest month of the year, averaging only 1.9" of rainfall. By fall, cold fronts finally reach North Texas bringing welcome relief from the heat and another wave of showers and thunderstorms. October is ordinarily our second-wettest month, averaging 4.22" of rainfall.

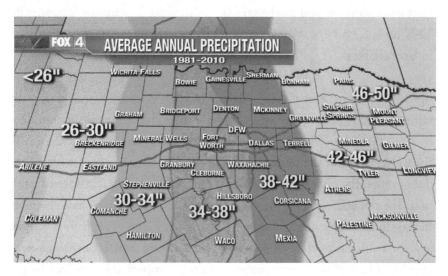

NORTH TEXAS RAINFALL DISPARITY

What is striking about precipitation in North Texas is the disparity in the annual average as you head from west to east. Generally speaking, using the I-35 corridor as a dividing line, the climate is progressively drier as you head west and progressively wetter as you head east. The difference in how much rainfall each region receives is reflected in the appearance of the landscape. Western portions of North Texas are generally more barren and feature foliage and vegetation that demand less water. Common specimens include Post Oaks, Cedar Elms, Juniper, Mesquite, and prairie grasses. Far eastern portions of North Texas are characterized by dense forest land. In addition to oaks, you will also find more water-demanding trees like Pecans, Walnuts, Loblolly Pines, and flowering Dogwoods.

The rainfall imbalance has a great deal to do with the source regions for our prevailing southerly wind flow. South winds entering western portions of North Texas blow in from the more arid regions of Northern Mexico. In contrast, southerly breezes blowing into eastern portions of North Texas originate over the Gulf of Mexico, where they have had the opportunity to tap into rich moisture. As meteorologists, we often observe cold fronts pass through dry until they reach the I-35 corridor. Showers and thunderstorms begin to develop in the Metroplex and then ignite as they encounter more Gulf moisture moving east of I-35. As you look at a map of annual precipitation for North Texas, the difference is glaring. Breckenridge, about 90 miles west of Fort Worth, averages about 27" per year, roughly 9" less than the annual average of 36.14". Emory, approximately 70 miles east of Dallas, averages 44" per year, or 17" more than Breckenridge. We have certainly made the case for the important role the low-level southerly winds play in North Texas weather. The weather we observe at the surface represents only a tiny fraction of what is going on in the earth's troposphere. While the depth of the troposphere varies with latitude, it generally extends from sea level up to about 35,000' over the United States, and is where nearly all weather takes place. The wind flow several miles above the ground is even more critical to determining changes in our local weather.

THE JET STREAM: WINDS ABOVE INFLUENCE WEATHER BELOW

Very swift-moving winds at high altitudes called "jet streams" form at the boundaries of warm and cold air masses. The core speeds of these jets can reach up to 250 miles an hour in winter. In February of 2019, a Virgin Atlantic flight from Los Angeles to London experienced that phenomenon firsthand. In a tweet that went viral, pilot Peter James wrote, "Almost 800 miles an hour now. Never ever seen this kind of tailwind in my life as a commercial pilot!" Getting swept up in a belt of strong jet stream winds over the eastern United States enabled the Boeing 787-9 Dreamliner to reach 801 miles an hour, which equated to a 240-mile-an-hour bump over its traditional cruising speed of 561 miles an hour. Thanks to its phenomenal pace, the flight arrived forty-eight minutes before its projected arrival.

The polar jet stream follows the boundary between the warmer middle latitudes and frigid polar latitudes. The subtropical jet stream is found much farther south, where tropical and mid-latitude air meet. When the north-to-south temperature contrast is greater (usually during winter), these "rivers of air" are stronger. A river accurately illustrates this flow because the stronger winds are in the center and weaken as you approach the river's "banks," or outer edges. The rotation of the earth directs these winds to travel on average from west to east, but the flow often meanders north to south in a wavelike pattern, with troughs and ridges forming along its course. The troughs mark dips in the jet stream that allow polar air to move southward toward the Equator. Conversely, peaks or ridges allow warmer air from the Tropics to move north toward the poles.

The position of these jet streams tend to correspond with the sun's position in the sky. As the sun's elevation in the sky increases during the warmer months, the latitude of the jet stream shifts northward. By summer, the jet stream is typically found near the US-Canadian border. As the sun's elevation decreases during the cooler months, the jet stream shifts southward again. These jet streams are significant because they can promote the development of mid-latitude storms and steer these weather systems across the country. Therefore, they drive the weather year-round in the Northern Hemisphere.

IT ALL COMES TOGETHER IN TORNADO ALLEY

In Tornado Alley, extending from Central Texas to South Dakota, a complicated choreography frequently takes place between the jet stream, the dry line, and these contrasting air masses. One additional partner that can join in the atmospheric dance is an area of low-pressure that develops just east of the Rockies. These "lee-side lows" commonly form when a westerly oriented jet stream blows over the Rockies. As these winds descend the eastern slopes, the air is stretched vertically. This enhances vorticity (whirling motion of air) in the atmosphere and "spins up" an area of low pressure. The process is complicated, but the movement is similar to how ice skaters pull their arms in close to their bodies in order to spin faster.

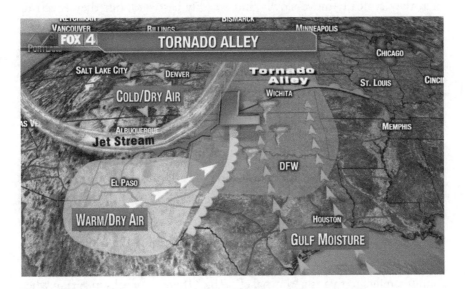

Lee-side lows typically form over southeastern Colorado or in the vicinity of the northern Texas Panhandle. As the low-pressure center deepens, the pressure gradient strengthens between the lee-side low and high pressure over the southeastern United States, or Bermuda High. This pressure gradient creates strong, southerly winds that sweep Gulf moisture north into Kansas and Missouri. A warm front marks the leading edge of this warm, humid air. To the northwest, either cold, dry air or cool Pacific air pushes in behind a cold front. Finally, to the west, a third air mass comes into play. Air that has been heated over the high terrain of the Desert Southwest and northern Mexico tends to drift northeast across Texas. This warm, dry continental tropical (cT) air descends from the Southern Rockies and Mexican Plateau and moves into the High Plains of West Texas, serving as the impetus for a dry line to develop. The dry line marks the leading edge of this warm, dry air.

Dry lines are often a catalyst for strong to severe thunderstorms, especially in spring. While they are often compared to cold fronts, they are a completely different animal. Compared to that of fronts, the movement of dry lines is far less predictable. At times dry lines creep forward, and at others they jump sporadically. They also "slosh" back and forth. During the morning, they typically move slowly and steadily eastward, but they speed up during the afternoon. The eastward jump is tied to strong heating. As the intense West Texas sun heats up areas west of the dry line, the warming encourages "mixing" of the atmosphere. Strong westerly winds thousands of feet above the ground are mixed down to the surface and propel the dry line eastward. Sometimes "bulges" will develop in the dry line where a small segment of the boundary moves faster or where one segment retrogrades while another continues its eastward push. Storm chasers will often converge on these dry line bulges, a favored location for storm initiation. As the sun goes down, the lower atmosphere to the west of the dry line cools rapidly. The mixing comes to a halt and the winds die down, causing the dry line to retrograde to the west during the evening. Dr. Lee Grenci, my former Penn State meteorology professor known for his wit and catchphrases, refers to this as a "diurnal cha-cha." There are weeks in spring when this cha-cha repeats itself day after day, firing off repeat episodes of severe weather in North Texas.

One springtime "wild card" that can make or break a forecast and determine whether a storm chase goes boom or bust is often tied directly to what is called "the cap." Recall that the dry line develops as air is heated over the Mexican Plateau and high deserts of eastern New Mexico and is then blown northeastward by winds aloft. Since this air originates at elevations around 4,000' to 7,000', it tends to remain at a similar altitude as it is swept eastward. This plume of warm, dry air continues to spread east of the surface dry line and forms a cap. Air parcels rising into this layer become cooler than the surrounding air and lose their buoyancy, so the cap acts as a lid and suppresses the development of thunderstorms. Determining when and if rising air can break through the cap is crucial to the forecast. A weak cap that breaks early in the day when surface temperatures are cooler typically does not favor intense storms because of the limited energy to tap. On the other hand, a strong cap is often not overcome, and the result is a sky filled with "lazy cumulus" clouds. Therefore, a cap that is neither too weak nor too strong is most favorable for severe weather. A moderately strong cap can act like a pressure cooker by allowing instability to build throughout the day as temperatures continue to rise. When the cap is finally broken, tremendous energy can be released, allowing for strong updrafts and the development of explosive thunderstorms.

Today, meteorologists have a plethora of tools and technology at their disposal to diagnose pivotal factors such as the cap and determine if skies will turn from sunny to stormy. How were weather forecasts made before the era of computers, satellites, and radar? Who laid the foundation for the modern science of meteorology? Let's turn the clock back to June 1944, when the fate of the world may very well have been hanging on the weather forecast.

CHAPTER 3

WEATHER FORECASTING

TECHNOLOGY/COMPUTER MODELS

A rguably, the most important weather forecast ever made was for D-Day, June 6, 1944. Group Captain James Stagg of the British Royal Air Force led three teams of meteorologists from the Royal Navy, the British Meteorological Office, and the US Strategic and Tactical Air Force. They were assigned the daunting task of forecasting the weather conditions for the notoriously stormy English Channel during the Allied invasion of France. Supreme Commander General Dwight D. Eisenhower and his fellow Allied commanders had a three-day window to launch the greatest invasion in military history. From June 5 to June 7, a nearly full moon would illuminate potential obstacles and landing places for gliders at night, and a low tide would expose the elaborate underwater defenses installed by the Germans. Weather and visibility would make or break the massive operation involving 156,000 troops, nearly 2,500 aircraft and gliders, and 6,939 ships and amphibious vessels. Air operations required mostly clear skies so bombers could see their targets and paratroopers could spot their landing locations. Naval forces required light winds and relatively calm seas so landing craft did not capsize before reaching the beaches of Normandy.

American Troops Approaching Omaha Beach, June 1944 *(Public Domain)*

THE BIRTH OF MODERN WEATHER FORECASTING

Making such a complex forecast would be difficult enough using today's technology, let alone before the era of weather satellites and computer models. Weather data over the North Atlantic were scarce, so the Allies flew weather reconnaissance missions to gather additional observations. The Allies had also cracked the Germans' "Enigma" code, which enabled them to intercept weather reports from German u-boats. Stagg's teams of meteorologists were all well versed in the Polar Front Theory that an extraordinary group of scientists based in Bergen, Norway, had published just after World

War 1. The group was led by Vilhelm Bjerknes, a Norwegian meteorologist and physicist, and included his son Jakob, Carl-Gustaf Rossby, Halvor Solberg, and Tor Begeron, among others. Their school of thought, known as the Bergen School of Meteorology, gave the world a working model of a mid-latitude cyclone, or low-pressure system, progressing through the stages of birth, growth, and decay. An important part of the Norwegian Cyclone Model was the theory that these cyclones form along "convergence lines" where warm and cold air masses meet. The convergence line that separated cold and warm air to the west and north of a low-pressure center was dubbed a "squall line," while the "steering line" separated warm

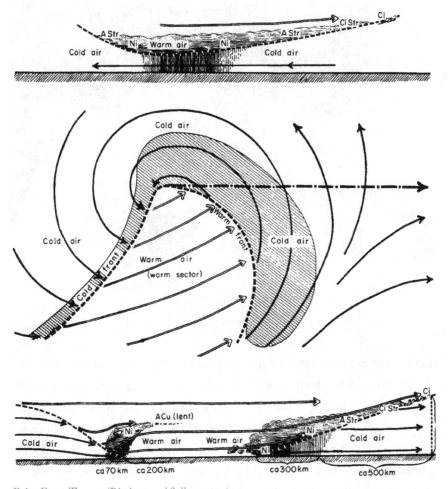

Polar Front Theory *(Bjerknes and Solberg, 1922)*

and cold air to the south and east of the low. Bjerknes and his colleagues later termed these lines the "cold front" and "warm front," an analogy to battle fronts in World War I. They also formulated a model of the cloud and precipitation patterns associated with a cyclone showing that rain or snow occurs on the cold side of the advancing warm front as the warm air glides up and over the large region of cold air. Where cold air advances to the rear of the storm, squalls and showers develop as the warm air is lifted abruptly and displaced. Their groundbreaking research formed the basis for much of modern weather forecasting.

Jacob Bjerknes *(Geophysical Institute, University of Bergen)*

Gathering measurements of air temperature, pressure, and humidity, the D-Day forecasters drew weather maps showing centers of high and low pressure, and located warm and cold fronts. The American and British meteorologists differed in their preferred method of forecasting and this led to heated disagreements between the teams. The American meteorologists preferred analog forecasting, which compared current weather analyses to historical weather patterns to find a case that looked similar. The problem with using analogs to make forecasts is that various weather features rarely align themselves the same way again, and even small differences between the past and present can drastically change the outcome. The British meteorologists preferred to use the weather analyses and their knowledge of the Polar Front Theory to determine how atmospheric conditions would change.

The difference in methodology led to sharp differences of opinion but, ultimately, the teams reached a consensus. On June 4, Group Captain James Stagg advised Eisenhower to delay the June 5 planned invasion. He reported that winds in the English Channel were likely to be force 5 on the Beaufort Scale, at roughly twenty to twenty-five miles an hour, making seas too rough for landing vessels. Worse, skies would be overcast, making paratroopers landing on their marks impossible and rendering precision bombing out of the question. Eisenhower took their advice and delayed D-Day by twenty-four hours. The weather for June 6 did not look ideal, but Stagg and his meteorologists believed winds would weaken and skies would clear enough to launch the attack. Meantime, German forecasters expected that white-capped seas and gale-force winds were likely to continue until mid-June. Thus, Nazi commanders believed that an Allied invasion was not imminent. Many, including German Field Marshal Erwin Rommel, left their coastal defenses. Despite pouring rain and howling winds on June 5, Eisenhower placed his faith in his forecasters and gave the go-ahead for D-Day on June 6. With the element of surprise on their side and weather that improved by midday, the successful Allied attack was a huge turning point in World War II.

EVERY FORECAST BEGINS WITH AN EYE TO THE SKY

Since that history-making forecast, weather technology has made huge leaps forward. Today, meteorologists have access to dual-polarization Doppler radar, high-resolution satellites, computer models, and automated surface-observing systems that enable us to see the atmosphere in motion like never before. Despite having these advanced tools at our fingertips, predicting the

TOP: Evan Andrews and Myself Watching Live Chaser Video *(KDFW-FOX 4 News)*
BOTTOM: With Ali Turiano during Severe Weather Coverage *(KDFW-FOX 4 News)*

weather still requires today's meteorologist to be a keen observer. For instance, many severe weather days hinge on breaking the cap, and there's no better way to tell if the cap is holding or if it's in the process of breaking than scanning the sky. If you see flat cumulus with no vertical growth, updrafts are struggling to overcome the cap. On the flip side, quickly growing towering cumulus, with their distinctive cauliflower tops, are a definite sign the cap is breaking and the prospects for severe weather may be increasing. Similarly, when forecasting fog and assessing the impacts of precipitation, you can't rely solely on computers and technology. While fog can be seen on satellite imagery, that imagery won't tell you how dense the fog is or to what degree it is restricting visibility. Likewise, knowing what type of precipitation is falling, how it is impacting the roads, the precise size and quantity of hailstones, and whether reported wind gusts are causing any damage can only be determined by sticking your head out the window.

Assessing the current weather situation entails examining satellite and radar imagery and combing through local and regional weather observations. Satellite animations can show approaching upper-air disturbances and fields of cumulus clouds where showers and storms may form in the coming hours. Radar can identify low-level boundaries, such as gust fronts, and disturbances generated by old thunderstorm complexes, called "meso-scale convective vortices," which can serve as catalysts for new storm development. Surface weather observations show temperatures, dew points, wind speed and direction, and air pressure tendencies, all of which are critical for identifying the location of warm fronts, cold fronts, dry lines, and high- and low-pressure centers. Finally, given all the current data you have just pored over, you need to ask yourself if yesterday's forecast verified. If it didn't, determining where it went wrong is critical to avoid repeating that mistake. But take it from this battle-scarred meteorologist—the next serving of humble pie is never too far away.

NUMERICAL WEATHER PREDICTION

When I was a child, I loved to build plastic model airplanes and construct bridges, Ferris wheels, and elevator lifts with an Erector set. These "physical models" are smaller, realistic representations of real-world objects that we can assemble to help us understand how the full-size versions work. But some things, like the weather, can't be imitated with plastic and metal pieces or pulleys from a kit. Instead, scientists use mathematical models to simulate the atmosphere, in what is known as "numerical weather prediction." These models allow us to understand complex systems using mathematical equations to forecast systems' future behavior. In the case of the weather, these atmospheric models solve dozens of complex equations based on the laws of physics to determine how things like horizontal and vertical air motions, temperature, pressure, and moisture will change with time.

Weather models utilize a system where forecast points are laid out in a grid over the area they cover. The more grid points there are in the model, the finer the detail in the resulting forecast. Think of a one-inch by one-inch square on your computer screen. The more pixels in that square, the smaller the pixels are, which results in a clearer and finer picture. Using more grid points, however, requires more calculations. Some models, such as the Global Forecast System (GFS), forecast the state of the atmosphere

384 hours (sixteen days) into the future. The GFS model requires over ten quadrillion (ten followed by *sixteen* zeros) calculations for a complete model-run forecast. The model runs four times daily, and each cycle takes two hours to complete.

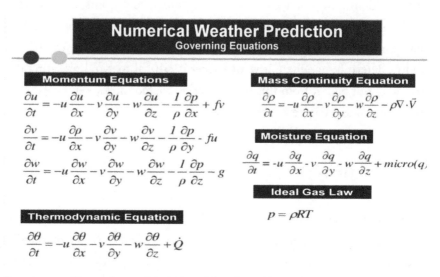

Numerical Weather Prediction
Governing Equations

Momentum Equations

$$\frac{\partial u}{\partial t} = -u\frac{\partial u}{\partial x} - v\frac{\partial u}{\partial y} - w\frac{\partial u}{\partial z} - \frac{1}{\rho}\frac{\partial p}{\partial x} + fv$$

$$\frac{\partial v}{\partial t} = -u\frac{\partial \rho}{\partial x} - v\frac{\partial v}{\partial y} - w\frac{\partial v}{\partial z} - \frac{1}{\rho}\frac{\partial p}{\partial y} - fu$$

$$\frac{\partial w}{\partial t} = -u\frac{\partial w}{\partial x} - v\frac{\partial w}{\partial y} - w\frac{\partial w}{\partial z} - \frac{1}{\rho}\frac{\partial p}{\partial z} - g$$

Mass Continuity Equation

$$\frac{\partial \rho}{\partial t} = -u\frac{\partial \rho}{\partial x} - v\frac{\partial \rho}{\partial y} - w\frac{\partial \rho}{\partial z} - \rho\nabla\cdot\vec{V}$$

Moisture Equation

$$\frac{\partial q}{\partial t} = -u\frac{\partial q}{\partial x} - v\frac{\partial q}{\partial y} - w\frac{\partial q}{\partial z} + micro(q,$$

Ideal Gas Law

$$p = \rho RT$$

Thermodynamic Equation

$$\frac{\partial \theta}{\partial t} = -u\frac{\partial \theta}{\partial x} - v\frac{\partial \theta}{\partial y} - w\frac{\partial \theta}{\partial z} + \dot{Q}$$

Equations used in weather models *(Adapted from Samuel Jackson)*

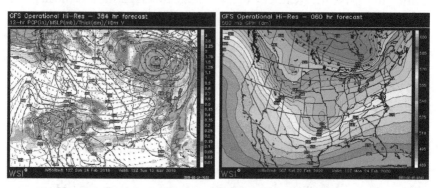

LEFT: 384-Hour GFS Precipitation Forecast *(The Weather Company)* RIGHT: 60-Hour GFS 500 mb Forecast *(The Weather Company)*

These models are fed with atmospheric and oceanic data gathered from all across the globe. Each day, billions of weather observations are taken from satellites, weather balloons, aircraft, ocean buoys, and ground observing stations. These measurements of air and sea temperature, air pressure, moisture, and wind speed constitute a tidal wave of data describing the atmosphere's current state, and this information is plugged into the equations. Computer programs are then used to solve the equations in order to calculate what the atmosphere will look like a few minutes into the future. Those results are fed into the same equations again to predict a few more minutes ahead in time. This is done repeatedly over a series of short time increments for each grid point and for as many as 128 levels in the atmosphere to help predict the formation, intensity, and track of weather systems. Generating these forecasts for multiple computer models requires some serious number crunching. Meet "Luna" and "Surge." No, they are not two participants about to smackdown in the WWE. These were the names given to the National Oceanic and Atmospheric Administration's (NOAA) newest twin supercomputers, each about the size of a school bus. These massive computers are equipped with fifty thousand processors and can perform up to three quadrillion calculations per second. NOAA's combined weather and climate supercomputing system has the ability to process eight quadrillion calculations per second.

NOAA Supercomputer *(Orlando, Florida)*

The arsenal of clubs a golfer carries in their bag is impressive. The typical "weekend warrior" set includes twelve to fourteen clubs. Fairway woods and hybrids are key to playing long shots from the fairway and rough. Irons are designed to hit shots from specific distances. Wedges are used for shorter approach shots into the green and getting out of bunkers. Finally, putters are the most specialized club and usually the last one used to finish the hole. Just as each club serves a different purpose, each weather model is uniquely suited for different scales of weather and types of weather. Also, just as some golfers blame a bad game on their clubs, meteorologists have a tendency to pin a bad forecast on the computer models.

There are basically two categories of models; global models that cover the entire globe and make forecasts up to two weeks into the future and regional and mesoscale models that cover a particular region of the globe and make forecasts two to three days into the future. The two most widely used global models are the Global Forecast System (GFS) run by NOAA and the European run by the European Center for Medium-Range Weather Forecasting (ECMWF). The more regularly used regional/mesoscale models are the North American Mesoscale (NAM), the Weather Research and Forecasting (WRF), and the High Resolution Rapid Refresh (HRRR). The North American Mesoscale model is run by NOAA to generate weather forecasts over the North American continent at various horizontal resolutions. The twelve-kilometer NAM (with twelve-kilometer grid spacing) is run four times daily and produces forecasts eighty-four hours into the future, while a higher resolution three-kilometer version produces forecasts sixty hours out. The WRF model was developed jointly by the National Center for Atmospheric Research (NCAR) and NOAA and is designed to serve both atmospheric research and forecasting needs. The WRF serves as the basis for NOAA's three-kilometer-resolution HRRR model, which is updated hourly. You want more models? We've got them! The Geophysical Fluid Dynamics Laboratory

(GFDL) model and the Hurricane Weather Research and Forecasting (HWRF) model are both run by NOAA. The GFDL and HWRF are the only models that provide specific intensity forecasts of hurricanes. Canada also operates its Global Environmental Multiscale (GEM) model, and the Met Office, the United Kingdom's national weather service, operates the UKMET model. A variety of other models exist, all operated by different agencies, but they are less reliable and therefore not used as frequently as the others mentioned.

THE "BUTTERFLY EFFECT"

Back in 1972, MIT meteorology professor Edward Lorenz was speaking at the annual meeting of the American Association for the Advancement of Science. During his presentation, Lorenz posed the question, "Does the flap of a butterfly's wings in Brazil set off a tornado in Texas?" While Lorenz was embellishing to perk up ears in the audience, it was an effective means to illustrate his idea of "Chaos Theory." He had made the discovery ten years before his famous speech, while he was running a computer simulation in which he rounded one variable from .506127 to .506. Lorenz was astonished when the seemingly minor alteration of one number drastically changed the results. The unexpected outcome led Lorenz to a groundbreaking insight—even small differences in a dynamic system, such as the atmosphere, could trigger vast and unexpected consequences. The "small differences" Lorenz referred to occur every time a computer model is run. Why? First, it is impossible to gather a set of perfect weather observations to accurately describe the current state of the earth's atmosphere. Second, many complex processes are nearly impossible to replicate with any model. These include phenomena such as evaporation, condensation, and the myriad ways that the sun's radiation is absorbed and reflected by clouds, atmospheric gases, and the ground. This means numerical weather prediction will always include errors in the initial data fed into the computer models, and these errors can grow exponentially over time, leading to large errors in the forecast.

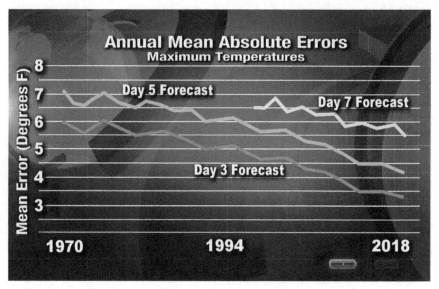

Adapted from NOAA *(Weather Prediction Center)*

The graph above shows mean absolute errors (either too high or too low) of the maximum temperature forecasts by the Weather Prediction Center, a branch of the National Weather Service that composes national forecasts for temperatures and precipitation. You will notice that forecasts have been steadily improving since 1970. The mean seven-day forecast error (yellow line) in 2018 was 5.5°F, which is significantly more accurate than the three-day forecast (red line) in the early seventies. Beyond seven days in the future, the accuracy of temperature forecasts really starts to suffer, and at nine or ten days the forecast error balloons to about seven degrees. At that range, forecasting is really no more accurate than using the climatological high temperature as a guide for how warm or cool it will be ten days from now. Imagine our dismay (speaking on behalf of all of my colleagues at FOX 4) when we see a post on social media accompanied by a ten-day computer model forecast indicating several inches of snow for North Texas. Unfortunately, it happens every winter. At that time horizon, it may be a disturbance over six thousand miles away in the Sea of Japan that one computer model is showing to be the catalyst for the snowstorm. Could it happen? While it's not out of the realm of possibilities, there's a much better chance of it being something entirely different, ranging from 38°F and rain to 70°F and sunshine. Yet the "snowpocalypse" post goes viral, leaving the FOX 4 weather team to explain why we believe it is far too early to predict a crippling snowstorm.

ENSEMBLE FORECASTS

The only way to achieve a 100% accurate forecast would require a perfect computer model to run with a set of observations at literally every single point in the atmosphere. That is simply not possible. This is why ensemble forecasts can be very useful, especially when we are faced with a challenging weather event shrouded in uncertainty. Ensemble forecasts run the same weather model multiple times from slightly different initial conditions. For example, a small adjustment may be made to the initial temperatures, and the model run again. Next, the initial pressure observations may be tweaked, and the model run a third time, and so on. This gives forecasters a range of different possible forecast outcomes. The complete set of forecasts is referred to as an ensemble, and the individual forecasts within it are ensemble members.

You have likely seen us air an ensemble forecast in our weathercasts during hurricane season to show the different projected hurricane tracks. Many meteorologists call this graphical depiction of an ensemble forecast a "spaghetti plot." Plotted on a map, each of those squiggly lines, which look like spaghetti, represents an individual ensemble member. A considerable spread in the members translates into uncertainty about the future track. A tight clustering of members indicates confidence in where the hurricane will make landfall.

A forecaster may also use ensembles to make rainfall or snowfall projections, or to show the development of an upper-level ridge or low-pressure trough. The ensemble version of the European model is known as the Ensemble Prediction System (EPS). The EPS generates fifty-one separate forecasts or ensemble members. The Global Ensemble Forecast System (GEFS) is the ensemble version of the GFS model. The GEFS is made up of twenty-one separate forecasts, or ensemble members.

UNDERSTANDING A COMPUTER MODEL'S LIMITATIONS

Every computer model has strengths, weaknesses, and biases. Mesoscale models, like the NAM, are helpful tools when forecasting thunderstorms, while global models like the GFS and ECMWF are more adept at handling larger storms like hurricanes and nor'easters. Here in North Texas, all of the models have difficulty capturing the timing and intensity of Arctic air masses. Many models tend to be too slow predicting the arrival of precipitation, particularly with a strong surge of warm air pushing in from the south (warm advection). Computer models also struggle mightily at times with predicting thunderstorm development, particularly on days when a cap is present, or in the absence of a well-defined lifting mechanism like a cold front or dry line. Even high-resolution models with grid spacing set at three kilometers cannot resolve the swirling eddies and rising motion of air and moisture that are key to cloud and thunderstorm formation, as these often occur on a scale less than one kilometer. Thus, a meteorologist must lean on his or her knowledge, experience, and intuition when forecasting.

My long-time colleague Evan Andrews believes weather forecasting is a blend of science and art. "The science part is black and white. It's just data and numbers. But looking at the atmosphere using several different computer models turns forecasting into abstract art. Is the data believable? Which one looks more like you would expect?" As meteorologists, we need to do the "quality control" on each model run. If the three- to six-hour forecasts are not accurately depicting the weather, chances are the longer-range forecasts are going to get progressively worse. In baseball terms, if your starting pitcher is getting shelled in the first inning, it's probably not his night, and you'd better make a call to the bullpen.

I have always firmly believed in a "consensus" forecast that combines information gleaned from all of the models. For instance, if four computer-model projections for rainfall at DFW are 0.75", 0.95", 1.65", and 2.25", perhaps forecasting 1.40", the average of the four models, makes the most sense. I consider other factors before rendering my forecast decision. Do all of the models have a good grasp of the current weather situation? If not, there may be one or two projections I toss out. Are the models forecasting the lower totals (0.75" and 0.95") trending up significantly from their previous runs? If so, that gives more credence to the higher totals and may convince me that 1.50" to 2.00" of rainfall is the more realistic outcome.

THE GFS-FV3: THE NEW AND IMPROVED GFS?

The European model has historically outperformed its American counterpart due to its higher resolution and its incorporation of real-time weather data to start the model. The initial data input into the GFS model is taken at a single time, whereas the "Euro" uses a twelve-hour period of continuous data compiled from satellites and weather stations. That gives the Euro a better sense of the current state of the atmosphere and a head start right out of the gate compared to the GFS. The exceptional performance of the Euro made headlines in 2012, when it was the only model to predict the left hook that brought Hurricane Sandy ashore in New Jersey. All the other models, including the GFS, had Sandy staying well out to sea. Backlash from Sandy got the attention of US policymakers, and additional funding was granted by Congress for NOAA to improve its model capabilities. In 2019, with the release of the GFS-FV3 (Finite-Volume Cubed-Sphere), NOAA took the first step toward improving its workhorse model. The higher-resolution model is able to simulate vertical movements such as updrafts, a key component of severe weather, at high resolution. In FV3, the microphysical processes, such as condensation, evaporation, freezing, and the growth of water droplets and ice particles, are also superior to the previous GFS, so the new model will likely be better at predicting where and how much rain will fall. New models, however, similarly to new software, always have some bugs to work out. The new FV3 model had an apparent "cold bias in the lower atmosphere," which caused it to dramatically overpredict snow amounts along the busy northeastern corridor between Boston, Massachusetts, and Washington, DC. The general consensus, though, is that the FV3 is performing better overall than its predecessor and should help meteorologists make better forecasts in the future.

THIRTY-PERCENT CHANCE OF RAIN WITH A SEVENTY-PERCENT CHANCE OF CONFUSION

One of the biggest challenges meteorologists face, outside of making an accurate forecast, is properly communicating the outcome to the public in a way that makes sense. On some days the forecast is easy to convey, "Tomorrow will be sunny and warm with a high temperature of 80°F." This is straightforward and everyone knows exactly what to expect. Introduce a chance of rain in the forecast, however, and the message can be lost in translation. "Tomorrow will be cloudy with a 30% chance of rain." Some people see that forecast and expect it to rain 30% of the time tomorrow. Others think it is going to rain over 30% of the area. Both of those interpretations are incorrect. The probability of precipitation (PoP), expressed as a percentage, describes the chance of measurable precipitation occurring at any point you select in the area. That 30% chance does not specify how much rain you will get. It just means there is a 30% chance of measurable rain, which is .01" or more. That may be a two-minute shower or two hours of heavy downpours. Also, a high chance of precipitation, say 70% or greater, does not necessarily spell a washout. Take, for instance, a nearly solid line of thunderstorms that is expected to move across North Texas. While chances are high that it will rain where you live, the rain may only last for a single hour of the day.

In addition to using percentages, meteorologists often use descriptive terms like "isolated," "scattered," and "likely" to indicate the chance of precipitation. When the forecast calls for "isolated showers" or "isolated thunderstorms," this implies a 20% chance or less of seeing a shower or thunderstorm. When the chances are in the 30% to 50% range, the term "scattered showers" is used. When the forecast calls for a 60% chance or higher, the term "likely" is used.

SEVERE WEATHER: LOW-PROBABILITY, HIGH-IMPACT EVENTS

Prior to issuing severe weather watches for the United States, the Storm Prediction Center (SPC) in Norman, Oklahoma, typically highlights the potential for severe weather using what are called "convective outlooks," which use a hierarchical number and category scale to delineate the overall severe weather threat. The scale is as follows: General Thunderstorms (light green), 1: Marginal Risk (dark green), 2: Slight Risk (yellow), 3: Enhanced Risk (orange), 4: Moderate Risk (red), and 5: High Risk (magenta). In general, the lower-number categories, Marginal and Slight, imply that severe weather is not expected to be widespread. Nonetheless, isolated tornadoes and baseball-sized hail are still possible on these lower-threat days. The upper-tier categories (three and higher) delineate areas where severe weather is expected to be more widespread. The map below shows the convective outlook issued on May 20, 2019. Note that areas in West Texas and portions of Western and Central Oklahoma were placed under a High Risk, which is typically issued only a few times per year. SPC forecasters issue a High Risk when they believe conditions are favorable for a major severe weather outbreak, and include the potential of long-lived and/or intense tornadoes. Fortunately, the May 20 event only produced one EF-3 tornado.

These outlooks, like the chances of rain, are based on percentages that are poorly understood by the general public. The misinterpretation likely stems from the public not realizing that the overall chance of a thunderstorm is typically much higher than the chance of severe weather. Meteorologists create graphics to visually explain these concepts and help viewers better understand the weather forecast. The graphic that viewers are most familiar with is the Seven-Day Forecast. On that graphic, we will label the chance of rain for the days we expect rain and/or thunderstorms. A problem arises on severe weather days when viewers equate that chance of rain—say it's 70%—with the chance of severe thunderstorms, when in reality that is simply not the case. Consider the "Probability in Your County" graphic below. This was a graphic we started airing during the 2019 spring severe-weather season in our FOX 4 weather segments to try to clear up the confusion. The "rain probability" corresponds to the overall chance of rain and is also the probability you will see labeled on the Seven-Day Forecast. It can come from showers or thunderstorms. Note, the probabilities of high wind, hail, and tornadoes are much lower than the probability of rain, as most of the thunderstorms we see on severe-weather days do not produce severe weather. They produce heavy downpours, frequent lightning, gusty winds below fifty miles an hour, and small hail the size of dimes and nickels. Frankly, meteorologists need to do a better job of communicating this distinction to viewers.

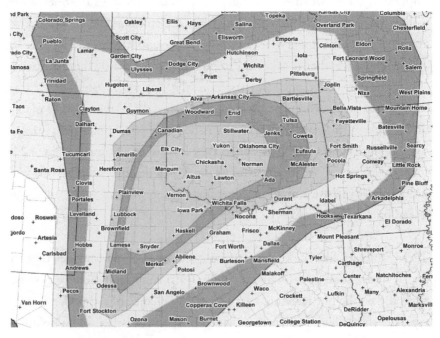

5/20/19 Severe Weather Outlook (SPC)

Since the majority of severe-weather risk days are on the low end of the threat scale, let's examine what a Slight Risk means on a Day One Outlook, which assesses the severe weather risk for the upcoming twenty-four hours.

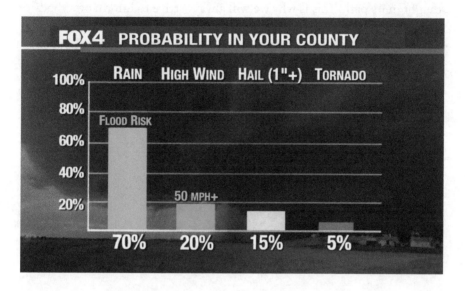

We'll assume all of North Texas is under a Slight Risk. According to the probabilities assigned to the Slight Risk category, this means there is a 5% chance of a tornado within twenty-five miles of your location, and a 15% chance of a severe wind gust (fifty-eight miles an hour or greater) or large hail (1" or larger) within twenty-five miles of your location. If you were to step out your front door and with the aid of a compass go twenty-five miles in every direction, the circular area covered would represent an area of roughly two thousand square miles, or the equivalent of Dallas and Tarrant counties combined. So, on a Slight Risk day, we can expect a 5% chance of a tornado and a 15% chance of either a severe wind gust or large hail somewhere in Dallas and Tarrant counties; the chances for any one particular neighborhood are even lower than that. You may be wondering why the National Weather Service and media outlets make a fuss over such a low probability event. The reason is twofold: First, while the chances of a tornado, damaging winds, or destructive hail are low for any one location, the potential impact to life and property is high. Second, we have no way of knowing which neighborhoods are going to get hit until the storms have already developed and may be only minutes away. As we have witnessed many times over the years, severe weather can destroy homes and businesses, toss vehicles around like toys, and fatally injure anyone caught in its path. This is why we will always strive to inform everybody of a potential threat.

CHAPTER 4

THUNDERSTORMS, SUPERCELLS, AND TORNADOGENESIS

L ieutenant Colonel William Rankin was a US Marine Corps pilot and veteran of both World War II and the Korean War, but he is best remembered as the only person to have fallen through a thunderstorm and lived to tell about it. His incredible story took place on July 26, 1959, as he and his wingman, Navy Lieutenant Herbert Nolan, were flying a pair of F-8 Crusaders along the North Carolina coast. They had climbed to 47,000' to stay above some strong thunderstorms. Shortly before their descent, Rankin heard a loud bump and rumble from the engine, which suddenly quit. Warning lights began to flash. He radioed Nolan: "Power failure. May have to eject." When he pulled the lever to deploy auxiliary power, the lever broke off in his hand. Despite Rankin's best efforts to stay aloft, the crippled aircraft began to nosedive. He was nine miles above the earth and had no pressure suit. He knew that he was in dire straits. Despite facing frigid temperatures of 58°F below zero and extremely thin air with mere trace levels of oxygen, he had no choice but to eject. He pulled the overhead ejection handles and shot himself into the atmosphere.

An emergency oxygen supply enabled Rankin to breathe, but he suffered instant frostbite and severe decompression that caused his abdomen to swell and blood to leak from his eyes, nose, ears, and mouth. In an August 1959 *Time* magazine article, he is quoted as saying, "It was a shocking cold all over. My ankles and wrists began to burn as though somebody had put dry ice on my skin. My left hand went numb. I lost that glove when I went out."

To make matters worse, his parachute deployed prematurely, forcing him to endure the perils of lightning, hail, and turbulence. "Boy, do I remember that lightning," Rankin told *Time* magazine. "I remember falling through hail, and that worried me. I was afraid the hail would tear the chute. Sometimes, I was falling through heavy water. I'd take a breath and breathe in a mouthful of water. I was blown up and down as much as 6,000' at a time. It went on for a long time, like being on a very fast elevator with strong blasts of compressed air hitting you."

After a grueling, painful ride that seemed unending, he finally exited the most intense portion of the storm. The temperatures were now warmer, the torrents of rain had tapered to light showers, and his parachute had somehow withstood the onslaught of wind and hail. He just needed to make a safe landing. However, the storm was not done with the marine pilot. A sudden gust of wind blew him into some trees, causing the chute to become snared in the branches and hurling Rankin into the trunk below. Rankin took a few moments to collect himself and check his watch. It was 6:40 p.m. Miraculously, he had survived a tumultuous forty-minute fall through a raging thunderstorm and had sustained only minor injuries. With his feet now back on firm ground, Rankin found his way to a backcountry road, where he was able to wave down a passing car. The driver took him to a nearby store in Ahoskie, North Carolina. From there, the pilot was ferried to the hospital, where he made a full recovery in a matter of weeks.

Rankin's incredible survival story provides a unique perspective of what happens inside a thunderstorm. How do these storms evolve from fair-weather cumulus clouds to towering behemoths of nature?

CLOUD FORMATION: DIRTY AIR PLAYS A ROLE

Consider a parcel or pocket of warm, humid air. Air pressure decreases with height in the atmosphere, so as the parcel rises, it encounters lower pressure. The lower pressure allows the air molecules to expand, leading the air to cool.

You can demonstrate how air cools as it expands by simply breathing out through your mouth. First, put your hand in front of your mouth and, with your mouth wide open, blow toward your hand. Now, with your lips puckered, allowing only a small opening in your mouth, blow toward your hand. In the second case, your breath is much cooler because the air inside your mouth is in a more confined space, and when it is released, it is allowed to expand.

Similarly, as the air rises, it continues to expand and cool. With sufficient cooling, it becomes saturated and the water vapor condenses, although this alone is not enough to create clouds. Our "dirty air" plays a vital role in the process. In the atmosphere, aerosols (microscopic particles of dust, dirt, smoke, and sea salt) attract water molecules. Called "cloud condensation nuclei," these tiny particles serve as the "seeds" upon which water vapor condenses and forms cloud droplets. When enough cloud droplets form, they become a visible cloud.

CLASSIFICATION OF CLOUDS

British pharmacist and naturalist Luke Howard, who had a keen eye for the sky and a fascination with weather, is considered the father of our modern cloud classification system. In his *Essay on the Modifications of Clouds* (1803),

Howard introduced an innovative system that employed Latin descriptions of clouds as they appear to an observer on the ground. He named high-altitude, wispy clouds "cirrus," which is the Latin word for "curl of hair." Cirrus clouds typically arrive in advance of low-pressure systems and hurricanes. He described puffy clouds as "convex or conical heaps, increasing upward from a horizontal base." Those he named "cumulus," meaning "heap" in Latin. Cumulus clouds indicate vertical motion taking place in the atmosphere. The base of cumulus clouds are generally flat and mark the altitude where moist, rising air condenses. Howard characterized the lowest clouds as "a continuous, horizontal sheet, increasing from below" and dubbed them "stratus," from the Latin word for "layer." Stratus clouds are thick, heavy-looking gray clouds that dominate the sky and block out sunlight. Precipitation normally does not occur with stratus clouds, although they may produce drizzle or mist. Howard also designated a precipitation cloud category, "nimbus," the Latin word for "rain." Other cloud names combine these four basic types. For example, "nimbostratus" refers to dark, low-level clouds that typically bring light to moderate rain, snow, or sleet, whereas "cumulonimbus" denotes a dense, towering cloud that often brings heavy precipitation.

THUNDERSTORM INGREDIENTS

Three basic ingredients are required for a thunderstorm to develop: moisture, instability, and a lifting mechanism to provide the "nudge." Moisture is to thunderstorms as gasoline is to a car. Without it, clouds cannot form, let alone develop into thunderstorms. Instability in the atmosphere allows air to continue to rise once it is given a nudge. An unstable atmosphere is one with warm, moist air near the ground and colder, drier air above this layer of warm air. Think of it as a hot air balloon. The hot air inside the balloon is less dense than the surrounding cool air, which makes it buoyant. In general, three things will increase instability: the sun's heating of the earth's surface, higher low-level moisture, and an environment in which temperatures get progressively colder with altitude.

Meteorologists love acronyms, and they use the acronym "CAPE" to quantify instability. CAPE stands for "Convective Available Potential Energy." It is a measure of instability through the depth of the atmosphere and can be thought of as the amount of fuel available to a developing thunderstorm. Higher values of CAPE mean the atmosphere is more unstable, favoring stronger thunderstorm updrafts and more intense storms. CAPE is determined by examining a vertical profile of temperatures and dew points (atmospheric sounding) obtained from weather balloons and is expressed in energy per unit mass (Joules per kilogram, or J/kg). In general, CAPE values less than 1,000 J/kg represent weak instability, 1,000 to 2,500 J/kg moderate instability, 2,500 to 4,000 J/kg strong instability, and greater than 4,000 J/kg extreme instability.

Even warm, moist air in an unstable atmosphere will usually not rise without something giving it an initial nudge or boost; some form of lifting mechanism is needed. That lift can be provided by an approaching cold front or dry line, heating from the sun, a low-pressure system, or terrain that forces the air to rise. Once the parcels of air are given a boost, they can continue to rise due to their buoyancy. When moisture, instability, and something to lift the air are present, the atmosphere is primed for thunderstorms.

THE LIFE CYCLE OF A THUNDERSTORM

Thunderstorms essentially go through three stages of development, beginning with the cumulus stage. A lone "fair-weather" cumulus cloud is thrust upward by a rising column of air called an updraft. Often, these first cumulus clouds face their demise as they entrain (draw in) drier air and their cloud droplets evaporate. These "pioneer clouds" pave the way for new cumulus to develop as they benefit from water vapor that evaporated as the first clouds dissipate. With sufficient moisture and lift, more of these cauliflower-shaped clouds build into "towering cumulus." These growing clouds now reach heights of 10,000' or higher and are dominated by updrafts. At this stage, small raindrops may begin to form, although the rising air keeps them suspended in the clouds. The updrafts continue to strengthen, and the churning turbulence causes the raindrops to collide and grow into larger droplets. Soon, the blossoming thunderstorm may reach heights of over 40,000', where air temperatures are well below freezing and ice crystals dominate. Strong winds at this altitude will typically flatten the top into an anvil shape. This majestic formation has become a cumulonimbus cloud.

Mature Stage of Thunderstorm *(NWS JetStream)*

In the second phase, called the mature stage, the cloud contains both an updraft and a downdraft. The storm is now at full throttle, producing its heaviest rain and hail and frequent lightning. Downdrafts develop as falling rain drags air downward. The evaporation of raindrops can strengthen the downdraft by cooling the air and making it denser, thus increasing its downward acceleration. When the downdraft reaches the surface, it spreads out along the ground. The leading edge of this rain-cooled air is called a gust front or outflow boundary. These cool, gusty winds can be a welcome relief on a hot, summer day and may cause temperatures to fall ten or fifteen degrees in a matter of minutes.

Cumulonimbus Cloud at Dusk *(Nicholas A. Tonelli)*

As heavy precipitation continues to fall, the downdrafts begin to dominate and the storm enters its final phase, the dissipating stage. Since warm, moist air can no longer rise, cloud droplets can no longer form. Soon, the downdrafts choke off the supply of warm, humid air and the storm quickly dies.

TYPES OF THUNDERSTORMS: FROM SINGLE CELL TO SUPERCELL

While similar in their makeup, all storms are not created equal. The most common are single-cell thunderstorms with one updraft and one downdraft. Often called "popcorn" or "garden variety" storms, they are usually weak, short-lived storms that last between thirty and sixty minutes. They live and die by the heating of the sun and may produce brief, heavy rain and lightning.

Next is the multicell storm, which is a group of single-cell storms that continually generates new cells. One classic example is a squall line, where the storms are arranged in a line, often parallel to a cold front or dry line. The band of storms continues to generate new development along the leading edge of the rain-cooled air (the gust front) as the cool outflow undercuts the warm, moist air ahead of the line. A low-hanging, well-defined, wedge-shaped cloud formation called a "shelf cloud" occurs along the leading edge of the gust front. Squall lines are often accompanied by "squalls" of high wind and heavy rain and may persist for hours. They can be hundreds of miles long but are typically only ten to twenty miles wide.

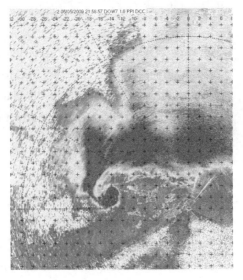

Hook Echo from DOW Radar *(Joshua Wurman)*

Shelf Cloud *(Credit: KairoK)*

The supercell, at the top of the thunderstorm hierarchy, is one of the most awe-inspiring sights in nature, and storm chasers and amateur weather observers alike refer to it as "the mothership." Unlike single-cell or even multi-cell storms, the supercell is driven by a persistent updraft that is tilted and rotating. It is typically two to six miles in diameter and can be present between twenty and sixty minutes before a tornado forms. Often, the storm will exhibit a distinctive hook echo on radar. More than 90% of supercells produce severe weather in the form of large hail, which can exceed the size of a grapefruit, and damaging winds. Furthermore, supercells are responsible for spawning most significant (EF-2 and stronger) tornadoes and virtually all violent (EF-4/EF-5) tornadoes.

A supercell resembles an atmospheric engine. The rotating updraft, called a mesocyclone, serves as the "fuel injector" that powers the storm, but there are many other parts that help drive it and define it. Let's take a closer look, with a top-to-bottom breakdown of the anatomy of the "mother of all thunderstorms."

ANATOMY OF A SUPERCELL

The anvil is composed of ice particles and forms the uppermost portion of the supercell. Rising air in the updraft loses its buoyancy as it encounters the warmer, more stable layer of air just above the tropopause (the boundary between the troposphere and stratosphere). Clouds stop developing and spread outward, forming an overhang that juts out in front of the storm. In some cases, the updraft builds up enough speed and momentum to break through the tropopause and form a dome at the top of the anvil. This is called an overshooting top. When you see one, it is a sure sign of a powerful updraft. Just below the anvil, you will often observe "mammatus" clouds—these pouch-like formations develop as cold air in the anvil sinks into the warmer air below. Although most commonly associated with thunderstorms, they can be found underlying other types of clouds. Supercells have two downdrafts, which serve as a sort of exhaust system. The forward flank downdraft (FFD) is associated with the main precipitation region where the heaviest rain and hail are falling. It is driven by rain-cooled air that cascades to the ground in the forward portion of the supercell. The rear flank downdraft (RFD) is generated as dry westerly winds in the mid-levels of the atmosphere run into the back side of the updraft. In essence, the storm blocks this mid-level wind flow and forces it downward. These dry mid-level winds cause some precipitation in the updraft to evaporate and cool the downdraft air, which gives it more negative buoyancy. As the RFD descends, it is compressed, which promotes considerable warming and

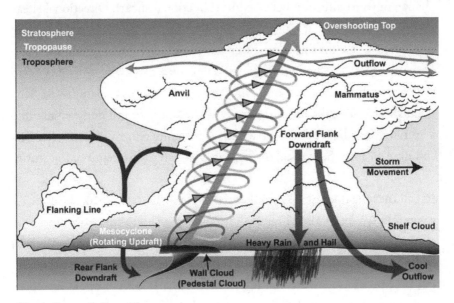

Classic Supercell *(Steve Horstmeyer)*

drying of the downdraft air. This can produce a small region of clearing in the cloud cover called a "clear slot." At the base of the storm, the wall cloud forms near the updraft/downdraft interface as the updraft ingests rain-cooled air from the downdraft region. This moist air quickly saturates as the updraft lifts it, forming an extension, or lowering, of the rain-free base of the storm. Wall clouds often generate excitement among storm chasers, but only a small percentage of wall clouds lead to tornado formation. Wall clouds that exhibit strong rotation or contain small, rapidly moving cloud fragments indicate that a tornado may be likely to develop. Finally, supercells typically have a series of smaller cloud towers to the south or southwest of the main storm tower. These smaller "stair-stepped" towers form the flanking line.

WIND SHEAR GETS A STORM ROTATING

Vertical wind shear separates a "run of the mill" thunderstorm from a long-lived, highly organized supercell. Vertical wind shear refers to either a dramatic increase in wind speed with height (speed shear) or a significant change in wind direction with height (directional shear). Directional shear produces rotation that can eventually spawn a tornado. Severe-weather days that are ripe for tornadoes often feature winds blowing from the south or southeast at the surface. The winds will turn clockwise with height so that, at roughly 4,000' to 5,000', they are blowing from the southwest. This wind shear causes the air in the lower atmosphere to roll horizontally like a spiraling football. Rising air within the thunderstorm updraft tilts this spiraling air into the vertical through at least half of the depth of the storm.

Recall that the updraft is not only rotating but tilted. In "weak shear" environments, where winds at higher elevations in the atmosphere are not much stronger than those near the surface, the storm's updraft is vertically aligned, and rain and hail fall back down through the updraft. The drag of the falling precipitation and the sinking motion enhanced by evaporative cooling essentially convert the updraft into a downdraft. In essence, the storm chokes on its own rain-cooled air. With stronger shear, in which those elevated winds are significantly stronger than the surface winds, the updraft is tilted. Because of this, precipitation falls well outside of the updraft, so the updraft and the rain-cooled downdraft remain separated. This allows for a continuous, undisturbed feed of warm, moist air that enables the updraft to persist for hours.

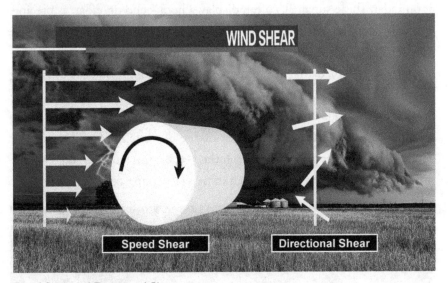

Speed Shear and Directional Shear

THE SUPERCELL SPECTRUM

Supercells can vary considerably in their appearance and the weather they produce depending on the amount of moisture available for the storm and the strength of high-altitude steering winds. In drier regions like the Texas Panhandle and West Texas, low-precipitation (LP) supercells are more common. The bases of LP supercells tend to be high above the ground due to dry air in the lower atmosphere, which requires air to be lifted to higher altitudes to reach condensation. Their cumulus updraft towers are typically heavily tilted with height due to winds markedly increasing with altitude. Due to their high cloud bases, LP supercells rarely spawn tornadoes, and most of the precipitation evaporates before reaching the surface. That said, powerful updrafts in LP supercells can support the production of very large hail, even when little or no rain is falling. LP supercells usually look more impressive in nature than they do on radar; their radar signatures are typically very weak due to the lack of precipitation they produce.

Classic supercells are harder to define, but you recognize one when you see it on radar. These supercells often display a hook echo on radar, as precipitation is wrapped around the mesocyclone due to strong counterclockwise rotation. Classic supercells produce all modes of severe weather, including very large hail, damaging straight-line winds, and tornadoes of varying intensity. Oftentimes, classic supercells will develop ahead of the dry line in western portions of North Texas such as Stephens, Eastland, and Young counties. As these storms move east of the dry line, they will encounter more Gulf moisture as they approach the I-35 corridor and undergo a transition from a classic to a high-precipitation (HP) supercell.

HP Supercell near Jolly, Texas *(Mike Mezeul II)*

HP supercells produce significantly more precipitation than their classic and LP counterparts. On radar, they typically don't display a characteristic hook echo and have more of a kidney bean shape. "Rain-wrapped tornadoes" can result from curtains of heavy rain being wrapped around by the rotating updraft, and these are nearly impossible to see approaching. Due to their high precipitation content, HP storms have dominant downdrafts that produce extensive cold pools of air at the surface that quickly cut off inflow into the storm. This tends to limit the duration of tornadoes, but these strong downdrafts favor gusty straight-line winds.

BIRTH OF A TORNADO

Now to address the million-dollar question: How do these Goliath storms spawn tornadoes? The process begins with the mesocyclone, or rotating updraft. The strength of the mesocyclone is tied to the degree of low-level wind shear and its orientation. A more robust mesocyclone will increase the chances of tornado development. Stronger wind shear produces more spinning of the air, which results in a more vigorous mesocyclone, as these "tubes of vorticity" are tilted vertically by the updraft. If the low-level wind shear is also aligned with the inflow into the storm (streamwise), then the air entering the updraft is spiraling like a football (see figure below) and the supercell is set up more favorably for tornado formation. On the other hand, if the low-level wind shear is not aligned, the air tumbles into the storm like a football kicked off a tee, and the disorganized flow that results is much less likely to lead to a tornado. This is only step one in the process; a strong mesocyclone alone cannot produce a tornado. In fact, many tornado warnings are issued based on impressive radar-detected rotation signatures that never result in tornadoes.

Dr. Paul Markowski, Professor of Meteorology at Penn State University and a leading expert on tornadogenesis, has been a principal investigator in several field studies to sample and probe the atmosphere in an effort to better understand how tornadoes develop. "We've known for a long time that to get tornadoes you need to have downdrafts as well as updrafts." Investigators have also determined that the temperature of the downdraft (the rain-cooled air that sinks toward the ground) also matters. "You need some coolness but not too much." Markowski relates the Goldilocks principle: "If it's too cool, the whole tornado formation process gets short-circuited because the cool air resists being converged and lifted due to its density." Tornadoes are more likely to form when the downdraft temperature is "just right." That tends to be just a few degrees cooler than the ambient air.

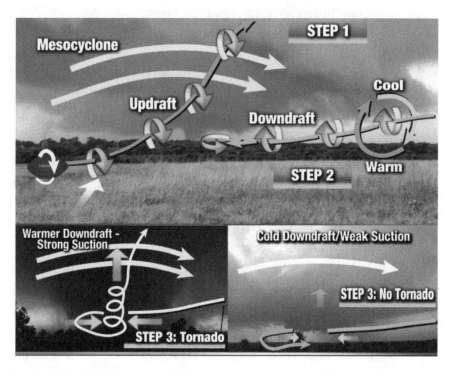

Adapted from Paul Markowski/Y.P. Richardson/William T. Reid

In step two of the tornado-development process, cool descending air parcels in the downdraft are bordered on the front side by rising warm air, and on the rear side by sinking cool air. This temperature and density gradient induces a spin that results in spiraling air parcels within the downdraft. Markowski compares this to pouring milk into a glass of water: "The horizontal spin you get on the flanks of a small pour of milk into a glass of water is the result of the same dynamics. The horizontal density difference between the heavier milk and lighter water generates horizontal spin."

As these spiraling parcels descend, they encounter a rising wind field surrounding the downdraft, which tilts this horizontal rotation into the vertical. This near-ground rotation, though, is very weak and nowhere near tornado strength. In 80% of supercells where this near-ground rotation is observed, a tornado fails to develop. Tornadogenesis requires one more critical step—the intensification of that near-ground rotation. This occurs primarily through the conservation of angular momentum. Strong mesocyclones generate upward-directed pressure-gradient forces (dynamic suction) that pull these spinning air parcels upward. Markowski explains this phenomenon: "The updraft is important for forcing angular momentum. The updraft can suck air upward, thereby increasing the spin the same way a figure skater increases their spin by converging mass to their center of rotation. This can amplify the vertical vorticity or spin by a factor of roughly 100." Thus, in step three, if the suction is sufficiently strong and the downdraft is just a few degrees cooler than the ambient air, a tornado is spawned.

With this understanding of how the mesocyclone and downdraft interact, meteorologists often look at two important factors to help them assess the risk for tornadoes: wind shear and relative humidity in the lowest 2,000' of the atmosphere. Strong, low-level wind shear promotes a stout mesocyclone that generates a lot of suction.

Higher relative humidity reduces the cooling effects of evaporation, so the spinning downdraft air is only slightly cooler than the surrounding air and is readily sucked upward by the intense updraft of the supercell.

Now that we have examined the inner workings of these powerful storms, chapter 5 will recount some large tornado outbreaks in North Texas over the past half century, beginning with a tornado that simultaneously captivated and terrified thousands of Dallas residents one afternoon in April 1957.

TORNADO OUTBREAKS IN NORTH TEXAS

)(

Twelve-year-old Gene Shoemake had been shopping with his older sister, Nancy, in downtown Dallas early in the afternoon of April 2, 1957. Gene and Nancy dropped by their father's workplace, a photography studio above the Majestic Theater, before boarding a bus to return home to Oak Cliff. Unbeknownst to the siblings and their fellow bus riders, a tornado had just touched down in southern Dallas County, near where present-day I-20 and Polk Street would be located. For the next fifteen minutes, the tornado seemed to skip and "bob and weave" as it moved north along Polk Street for six or seven miles. The first significant damage was reported around the Highway 67/Ledbetter area near Red Bird Airport (now Dallas Executive Airport). Meanwhile, the bus Gene and Nancy were riding on was heading westbound on Jefferson. They were virtually on a collision course with the tornado, which was gaining considerable strength as it moved into Oak Cliff. As it reached the intersection of Jefferson and Willomet, just a couple blocks south of the bus route, the children could see several homes had already had their roofs blown off.

1957 Dallas Tornado (us *Weather Bureau*)

Gene says that as the bus reached the corner of Jefferson and West 10th Street, the passengers looked out the windows and saw the tornado. "The tornado was barreling right toward us, destroying homes left and right. We saw lumber, bricks, trash cans, and other swirling debris flying through the air." The massive cloud of homes and fences and other belongings turned the afternoon sky completely dark. The bus driver, stricken with panic, stopped in the middle of the intersection. Other drivers did the same and fled their cars. They all feared the tornado would wipe them out. According to Gene, "The roar of the tornado was deafening and just as it approached us it lifted off the ground and went airborne over our bus and missed us." He watched, incredulous, as the fierce winds ripped a telephone pole out of the ground. "[The tornado] threw it completely through a new Ford automobile, in one side and out the other side, like an arrow had been shot through a tin can. Lumber had been shot through brick chimneys as if they were made of cardboard." The destruction all around them was incredible, but miraculously, everyone on the bus was spared.

Tornado Crossing Trinity River Levee *(US Weather Bureau)*

The tornado's onslaught ensued as it headed north into Kessler Park. It continued to ramp up, ripping roofs off homes near Colorado Boulevard and Winnetka Avenue, then crossing over what is now I-30. After traveling north for several miles, it took a jog to the northwest, crossed the Trinity River levee, and steamrolled into West Dallas. Now at its peak intensity and displaying multiple vortices, it was likely packing winds as strong as 175 miles an hour. Lillie Fuller, a pregnant thirty-one-year-old homemaker, was at her home on Vilbig Street. In a 1997 FOX 4 interview with Chip Mahaney, she recalls a friend screaming at her door, "Tornado, tornado!" Lillie looked out the window and immediately saw the tornado hurling debris in all directions. Houses, cars, and trees were being obliterated by the twister's powerful winds.

Fearing her home would be no match for the tornado, Lillie instinctively grabbed her keys, ran outside, and jumped in her car. She drove away but only made it to the corner of Vilbig and Homeland Street before her car's engine suddenly died. Lillie was stranded and thought she was a goner. As she sat in her car, paralyzed with fear, she witnessed the tornado tearing apart her home. She considers it a miracle that she escaped with her life and suffered no injuries. Several others would not be lucky enough to escape the tornado's fury. Among the casualties were three siblings, ranging in age from eighteen months to six years, who died when their Arlington Park apartment on Riverside Drive was leveled. Mercifully, the tornado began to lose steam after crossing Harry Hines Boulevard. By the time it reached the parking lot of Love Field it began "roping out," and then fully dissipated after crossing Bachman Lake.

Prior to the 2015 Sunnyvale-Garland-Rowlett tornado, the F3 tornado (EF-4 on the current Enhanced Fujita Scale) that hit Dallas shortly after 4 p.m. on April 2, 1957, was the deadliest in DFW history. During its fifteen-mile rampage over the course of roughly thirty-five minutes, it killed ten people. The Dallas tornado was one of twenty-eight twisters that tore across Texas and Oklahoma that day, and one of many outbreaks during the 1950s that claimed hundreds of lives. Just four years earlier, on May 11, 1953, an F5 tornado hit downtown Waco, killing 114 people. In fact, the Waco tornado outbreak, combined with the Flint-Worcester tornado outbreak that struck the Great Plains, the Midwest, and New England a month later, resulted in nearly four hundred fatalities and well over a thousand injuries.

The high fatality rates during the 1950s came at a time when severe weather prediction and detection were still in their infancy. The first computer models were being developed but were primitive and imprecise, and the US Weather Bureau's first weather radar system, the WSR-57, was not operational until 1959. The Dallas tornado was captured by a radar at Love Field used by air traffic controllers to help pilots land in poor visibility, but the images were low resolution and nothing like those shown during televised weathercasts today. Even still, some images were captured, and useful data was recorded. Amidst the devastation and tragic loss of human life, the Dallas tornado provided meteorologists with a wealth of valuable scientific data that advanced their understanding of wind fields and the structure of tornadoes.

SCIENTIFIC IMPORTANCE OF THE TORNADO

Unlike many tornadoes in North Texas that are shrouded by curtains of heavy rain and hail that make them very difficult to see, the 1957 tornado was set against cloudy skies void of heavy precipitation. This gave thousands of people in Dallas and Oak Cliff a remarkably clear view of the tornado along its entire path. Hundreds of photos were taken by as many as 125 people, and more than 2,000' of film of the tornado was shot. This provided meteorologists with the first photographic documentation of the entire life cycle of a tornado.

Scientists conducted exhaustive research over the next three years, studying every minute detail in the photos and every frame of the film. They analyzed how smoke from a smokestack interacted with the tornado and they tracked pieces of debris, frame by frame, to map the complex wind field and gather measurements of its rotational characteristics. The scientists also oversaw the illustration of a series of fifty-seven scaled sketches. The drawings were based on photos that depicted the size, height, and shape of the funnel from beginning to end. A comprehensive engineering study examined the structural failure of walls and roofs as well as the overturning of train cars to estimate the wind speeds. The US Weather Bureau combined all of the findings of Walter Hoecker, Robert Beebe, Dansy Williams, Jean Lee, Stuart Bigler, and E.P. Segner in a 175-page research report released in 1960.

Much of the information gleaned from that 1960 report, along with damage surveys conducted on the Fargo, North Dakota tornado in 1957 and the Lubbock, Texas tornado in 1970, served as the basis for the first tornado-intensity rating scale. Dr. T. Theodore Fujita of the University of Chicago, in collaboration with Allen Pearson, head of the National Severe Storms Forecast Center, introduced the Fujita-Pearson Scale in 1971. Better known as the Fujita Scale, it rated tornadoes from F0, with winds of forty to seventy-two miles an hour, to F5, with winds of 261 to 318 miles an hour. In 2007, scientists replaced the Fujita Scale with what they believed to be a more accurate Enhanced Fujita Scale. The EF Scale classifies tornadoes from EF-0 (sixty-five to eighty-five miles an hour) up to EF-5 (greater than two hundred miles an hour).

HISTORIC TREASURE DISCOVERED IN AN EAST TEXAS ATTIC

In my thirty-plus years in television news, I've learned that sometimes a good story will generate a lead to an even better one. Back in 2007, I reported on the fiftieth anniversary of the historic Dallas tornado. A few minutes after the feature aired on *FOX 4 News at Nine O'Clock*, I received a phone call from a woman from Wills Point named Vicki Lester. Vicki told me she believed she had some original 8 mm film of the tornado, shot by her late father, James Lester. She had discovered it in a thin tin canister labeled "Dallas Tornado" while rummaging through her attic. As the story goes, James had been working in Oak Cliff that day and had shot the footage. For thirty years, he held on to the film, until he passed away in 1987. For whatever reason, James never told his daughter about it. Vicki believes that, ironically, her father may have never seen the film he recorded because he didn't own a projector. Vicki unknowingly inherited the old film among her father's belongings. It wasn't until twelve years after her father's death that she discovered it while doing some spring cleaning in the attic. From there, the film got moved to the garage, where it sat on a workbench for another eight years.

We invited Vicki and her husband, Charles, to our downtown studios. We told them we would do whatever it took to find an 8 mm projector to enable them to watch the film. In the back of my mind, I thought the odds of the film surviving fifty years in a thin canister, exposed to extreme temperatures, were about the same as snow falling on the Fourth of July in North Texas!

Jim Reid, a FOX 4 colleague and television history buff, happened to own an 8 mm projector. He offered to bring it to the station and, just a few days after the phone call, Vicki and Charles met us at FOX 4 with the vintage film. About a dozen of us—meteorologist Ron Jackson, I, and several news managers and producers—gathered in a conference room where the projector and a screen were set up. The excitement was palpable as Jim loaded the ancient film into the projector. Were we about to witness an important piece of Dallas history, or would our hopes be dashed by the cruel course of time?

As the first images flickered to life on the screen, we were mesmerized. After fifty years with only a thin canister to protect and preserve it, the footage looked incredible. From his vantage point in Oak Cliff, James had captured several minutes of the tornado's life cycle, including the smaller vortices that danced around the main tornado. His footage showed the intense circulation narrowing into a "cone tornado" and then taking on the appearance of a skinny rope as it dissipated to the north over Bachman Lake. James had not stopped there; he traveled a few blocks over, into one of the heavily damaged neighborhoods, where he captured devastating images of people sifting through debris, homes with exterior walls leveled and roofs missing, and overturned cars. The film became the subject for a great follow-up story that aired on FOX 4 a few weeks later. James's "documentary" enabled thousands of North Texans to see the tornado and the wake of destruction left in its path. You can watch it by scanning the QR code here.

1957 Dallas Tornado Film (*KDFW–FOX 4 News*)

April 25, 1994 began as a typical spring day in Texas. A deck of clouds hung low in the sky as southerly breezes made for a mild, muggy morning. It was clear that the atmosphere was ripe for some form of severe weather later that day, with a strong, negative-tilted trough moving out of the Southern Rockies and swift "divergent" flow across Texas and Oklahoma. (Looking at a clock, using a noon-to-six-o'clock alignment as your neutral baseline, a one-to-seven-o'clock alignment is forward or "positive tilted," while an eleven to five alignment is "negative tilted.") The 500 mb chart showed the center or axis of the trough approaching Texas oriented from roughly ten o'clock to four o'clock. One branch of flow was moving from West Texas almost due north into western Kansas, while a diverging branch moved from West Texas to the northeast and east. In this situation, low-level air is forced to rise dramatically to replace the air at higher altitudes that is being swept away.

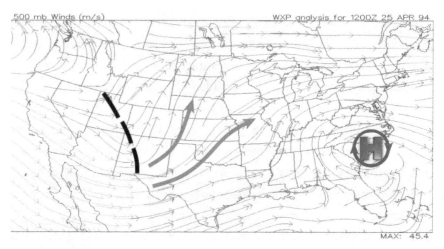

500mb Analysis on April 25, 1994 *(Plymouth State Weather Center)*

The morning overcast broke up during the afternoon, allowing the late April sun to heat the soupy air mass. Mid-morning storms developed along a dry line in West Texas but didn't reach North Texas until late in the afternoon. While much of this first round died out near sunset, a couple of supercells not only survived but strengthened as they rumbled into Tarrant County and headed for southern Dallas County. Tim Marshall, a pioneering storm chaser and frequent contributor to FOX 4 News, says he saw a lightning-illuminated tornado develop around 8:50 p.m. over Joe Pool Lake. It only lasted about two minutes before a large wall cloud formed to the east. He quickly lost sight of it after it became wrapped in rain.

Damage surveys the following day indicated that just after the storm crossed the lake and moved into DeSoto, it spawned a tornado that hit the newly constructed DeSoto City Hall. Moving east-southeast, the tornado tore roofs off several houses and then crossed I-35 into Lancaster. That's where the twister intensified. A one-block-by-three-block area of Lancaster was annihilated by the tornado's fierce winds. Many homes here were flattened and 80% of the historic town square was destroyed. In a 2004 interview, Sybil Roden and Virginia Coulston told me that the only portion of their home that had been left standing was the closet they were hiding in. Surviving the storm inspired them to write a poem they called "God's Closet."

The Lancaster tornado hugged the ground for approximately six miles and was up to a half mile wide. It damaged or destroyed 450 homes and fifty-eight businesses. The quick-hitting twister killed three people and injured forty-eight others, but it could have been much worse. Large hail was falling just north of the tornado on I-20. Many drivers stopped beneath bridges to shield their vehicles from the hail and, by doing so, created a roadblock that trapped hundreds of vehicles for miles behind them. Had the nighttime tornado turned north or even northeast toward the busy interstate, many more casualties could have resulted.

MARCH 28, 2000, FORT WORTH TORNADO

The views from the thirty-fifth floor of The Tower in Fort Worth are nothing short of spectacular, but early in the evening on March 28, 2000, that high perch in the heart of downtown offered a view of Mother Nature at her deadliest. Prior to its rebirth in 2005 as a high-rise luxury condo building, The Tower had been the Bank One Building and home to the Reata restaurant, located at the top of the building on the thirty-fifth floor. Mike Evans, co-owner and manager of the Reata at that time, was, coincidentally, hosting a group of meteorologists who were in Fort Worth for a weather conference. Evans says he received a call from a valet downstairs three minutes before the building was hit head-on. The caller told him a tornado was approaching from the west and heading across the Trinity River toward the nine-story Cash America Building. Without hesitation and with only seconds to spare, Evans and fifty of his restaurant staff herded roughly one hundred stunned restaurant patrons, including the meteorologists, into the concrete-reinforced interior stairwell. "It was like a bomb going off. The tornado's roar sounded like a train passing within 10 feet of us. The whole building shook and swayed." Evans says one gentleman wanted to go back into the restaurant to retrieve his laptop, but Evans refused to let him out of the stairwell. Within seconds, winds exceeding one hundred miles an hour and carrying all sorts of debris shot out all of the windows and blasted through the building. "It lasted about thirty seconds. When it was over, I immediately ordered everyone to go downstairs." Evans spent two hours surveying the damage. Most of the tables and chairs had been sucked or blown

out of the restaurant. "It was surreal. All of the windows were blown out. There were pieces of glass stuck in the walls. Had we not gotten that phone call from the valet parkers there would have been people killed up there."

2000 Fort Worth
Tornado *(Martin
Lisius/StormStock)*

Earlier in the day, the Storm Prediction Center had issued a "Moderate Risk" of severe weather as conditions were favorable for storms to produce large hail and damaging winds, but the tornado threat appeared to be mainly confined to the Red River Valley along a stalled frontal boundary. By mid-afternoon, low-level wind shear was increasing in the DFW area, raising the threat of tornadoes. At 4:38 p.m., a supercell was quickly forming in Parker County, prompting a severe thunderstorm warning. An hour later, the storm moved into Tarrant County, triggering another severe thunderstorm warning.

By six o'clock in the evening, radar signatures showed increasing rotation in the storm, and a hook echo was being detected. Storm spotters confirmed that this storm was growing better organized, and they were reporting a lowering wall cloud and baseball-sized hail falling north of the rotation along Loop 820 in North Fort Worth. Under such conditions, we need to be vigilant and carefully watch each scan of the radar. Meteorologists at the National Weather Service in Fort Worth were doing exactly that and, as a result, were able to spot winds converging just beneath the cloud base of the storm. They immediately recognized the imminent tornado threat and issued a warning for Tarrant County at 6:10 p.m.

At 6:18 p.m., a weak tornado touched down in River Oaks, gradually gaining strength as it moved southeast. The tornado plowed into a field house at Castleberry High School, catapulting debris, including rooftop air conditioning units, into the air. Continuing southeast, the twister crossed the Trinity River and plowed through the adjacent subdivision, damaging at least one hundred homes. At West Sixth Street and University Drive, the tornado turned to the east and slammed into the Linwood subdivision at 6:22 p.m., destroying several homes. The tornado's assault then turned deadly as it hit the Montgomery Ward distribution center, flipping over several parked trucks and causing a brick wall to collapse. A fifty-two-year-old man was killed while trying to reach shelter after warning others of the tornado, and a sixty-seven-year-old man was killed when the wall collapsed on him. These were the only fatalities attributed to the tornado.

The tornado crossed the Trinity River again at 6:24 p.m. and revved up to F3 intensity as it blew into downtown Fort Worth at 6:25 p.m.. The tornado pummeled several skyscrapers, including the Cash America building, Mallick Tower, the Bank One Tower, and the Union Pacific Resources building. Winds exceeding 145 miles an hour created a violent whirlwind of wood, bricks, sheet metal, glass shards, and office furniture that shattered thousands of windows. The Bank One Tower alone lost 80% of its three thousand windows. Thankfully, the downtown siege lasted only sixty seconds before the twister moved east of the central business district.

By 6:28 p.m., the tornado had dissipated near I-35W, but the parent storm kept chugging along I-30, east of Fort Worth, where it merged with a second storm that had developed earlier around Benbrook and moved northeast. The resulting "cell merger" became re-energized and spun up another tornado in Arlington around seven o'clock. The first reports of damage occurred

at a restaurant on South Cooper Street. Over the next fifteen minutes, the tornado gained strength as it moved east, roughly parallel to I-20, making for a harrowing commute for thousands still heading home. At least one eighteen-wheeler was overturned when the tornado crossed I-20 one mile west of Highway 360. The heaviest damage, rated F3, occurred near Bardin and Matlock. Near the end of the tornado's track, it produced F2 damage in a neighborhood just northeast of Grand Prairie Airport. Amazingly, there were no deaths or serious injuries in Arlington or Grand Prairie.

TOP: Fort Worth Tornado (Melanie and Robert Brownrigg) BOTTOM: Damage Inside the Mallick Tower (Melanie and Robert Brownrigg)

The late Al Moller, a well-respected meteorologist who worked for many years at the National Weather Service Office in Fort Worth, believed that storm interactions may have played an instrumental role in generating the Fort Worth tornado. Before the two storms merged, an outflow boundary from the Benbrook storm had moved north into

Fort Worth. That outflow boundary may have enhanced low-level wind shear and spinning air parcels (recall the spiraling football illustration). Moller also remarked that the outflow boundary was aligned parallel to the inflow entering the supercell. If that indeed was the case, those rotating air parcels could have easily been sucked into the updraft to spin up the tornado.

Another supercell produced a series of four tornadoes in rural areas thirty to forty miles southeast of Dallas later that evening, although these did not result in any injuries or significant damage. All told, there were nine tornadoes in North Texas that day. The Fort Worth and Arlington tornadoes caused nearly $500 million in damages. In addition to the two fatalities caused by the Fort Worth tornado, a third person was killed in North Fort Worth after being struck by baseball-sized hail. It was the first hail-related fatality in the US since 1979.

APRIL 3, 2012, TORNADO EMERGENCY FOR THE DFW AREA

As a meteorologist in North Texas, you know there will be challenging days forecasting and tracking severe weather—and then there are days like April 3, 2012. Early in the afternoon on April 2, two supercells with confirmed tornadoes tracked northward through Tarrant and Dallas counties, putting much of the Metroplex on high alert. Recognizing the grave nature of the situation, the National Weather Service issued a tornado emergency at 1:16 p.m. for Dallas County and a second at 1:25 p.m. for Tarrant County, each with very strong wording: "A tornado has been confirmed! This is an extremely dangerous and life-threatening situation. If you are in the path of this destructive tornado, take cover immediately."

The first tornado emergency was issued on May 3, 1999, in Oklahoma, when a powerful F5 tornado hit Bridge Creek, Moore, and Oklahoma City. These enhanced tornado warnings are issued when large or strong tornadoes affect highly populated areas and present a heightened danger of widespread damage and significant loss of life. It was the first time we recall witnessing two large tornadoes running up I-35 West and I-35 East together, threatening Dallas and Tarrant counties simultaneously.

At dawn, April 2 was relatively quiet and cool, but, south of DFW, tropical moisture surged north through the morning, as temperatures warmed into the eighties by noon. At the same time, an outflow boundary from morning storms in Oklahoma was moving south through North Texas. As it passed south of I-20, surface winds shifted to the east in the Metroplex, increasing the low-level wind shear, a key ingredient for enhancing rotation in storms. At noon, a squall line was beginning to take shape, from Parker County south to Bosque County, while two isolated thunderstorm cells popped up in Johnson and Ellis counties, along the outflow boundary. Innocuous at first, they blossomed quickly and began showing signs of rotation. Around half past noon, at the conclusion of *FOX 4 News at Noon*, the rotation in both storms began intensifying. Over the next several minutes, each radar update was more impressive than the last. The storm near Cleburne was initially the strongest. As reports of funnel clouds came into the FOX 4 Weather Center, we jumped on the air to update our viewers.

Covering an outbreak like this live on the air is best described as "controlled chaos." I was on camera in front of the green screen while Evan Andrews was 20' away in the Weather Center on another camera, tracking storms with our radar software, and weekend meteorologist Ron Jackson was monitoring the National Weather Service chat room. In this chat room, information comes in at a rapid-fire pace from emergency managers, city officials, trained storm spotters, police, EMS, and fire departments all across North Texas. We were also keeping an eye on our traffic cameras and live video feeds from our storm chasers.

We are trying to sift through a vast assortment of information to obtain what we deem to be the most reliable reports to confirm what we are seeing on radar. It is one thing to show our audience a hook echo or identify areas of strong rotation on radar, but what really resonates with viewers are photos and videos. In the case of this storm, it did not take long to get the compelling video we were looking to share, including the clip below.

The first tornado of the day occurred just north of Cleburne in Johnson County around 12:45 p.m. It damaged three manufactured homes and overturned a horse trailer. By one o'clock in the afternoon, the supercell to its east in Ellis County had produced a brief, weak tornado in Oak Leaf, in Ellis County. Over the next ten minutes, as the pair of supercells moved into southern Tarrant and southern Dallas counties, the situation escalated quickly. The eastern supercell was now producing a large tornado with debris and had moved into Lancaster.

Lancaster Tornado
(Steven Butler)

This prompted the National Weather Service to issue its first tornado emergency. Just a few minutes after the tornado emergency was issued for Dallas County, we saw live video of fifteen-thousand-pound semi-trailers being tossed into the air like toys at a trucking facility on 1-20 south of Dallas. Long-time and rock-steady FOX 4 reporter Shaun Rabb, who knows that area of Dallas County well, joined our team coverage, calling out familiar street names and landmarks to help pinpoint the exact location for our viewers. Five minutes later, one of our Tarrant County traffic cameras captured the large Kennedale tornado, with debris flying hundreds of feet into the air, as it was crossing Highway 287. Traffic had come to a dead stop, with some vehicles a mere fifty yards in front of the circulation. The traffic camera and video footage was a sobering reminder to our viewers of the power these tornadoes possess and it set the tone for our coverage.

By far, our biggest challenge was devoting equal time and coverage to our viewers in Dallas and Tarrant counties. It was agonizing to have to jump from one dangerous storm to the other, knowing that as you focused on one tornado, people in the path of the other tornado were being temporarily left in an information void. For the next hour, we juggled video clips, live traffic cameras, and live shots from reporters as we tracked these large tornadoes on radar from one neighborhood to the next. It was a team effort, with directors, producers, reporters, and managers all contributing.

Tornado Approaching Highway 287 in Kennedale *(Brad Rivers)*

It wasn't long after the Dallas and Tarrant County tornadoes dissipated that more tornadic supercells developed. One particular storm flared up over Mesquite around 3:15 p.m. and produced a brief tornado near the Mesquite Rodeo Arena. As the storm moved east into Kaufman County, it intensified as it approached Forney. Storm spotters confirmed that another tornado had touched down near downtown Forney. As we continued to track the area of rotation through Kaufman County, the damage reports we were receiving indicated this was a highly destructive tornado.

Minutes after the tornado hit the Diamond Creek subdivision, we shared live video from a storm chaser. Several homes in that neighborhood had been mangled, with only small portions of interior walls left intact. Your heart leaps into your throat seeing those live images. All you can do is hope and pray that people escaped without serious injuries and continue to warn others who are still in danger. Indeed, that same storm continued its rampage into Rockwall and Hunt counties, spawning another strong tornado in

Royse City, just before four o'clock in the afternoon. That tornado destroyed several homes, as well as a cabinet factory, and flipped over multiple vehicles.

This would turn out to be a marathon day that would continue late into the evening. A total of fifty-five severe thunderstorm warnings and eighteen tornado warnings were issued by the National Weather Service between 7:32 a.m. and 8:13 p.m. Over the next few days, survey crews confirmed twenty-two tornadoes, including EF-2s with estimated winds of 130 to 135 miles an hour in the Lancaster and Kennedale/Arlington areas, a high-end EF-3 packing winds up to 150 miles an hour in Forney, and an EF-2 in Royse City. The destruction across North Texas was staggering, with 1,100 homes damaged and another 349 homes completely destroyed, resulting in over $1 billion in damage. Yet the most important number of the entire event was "zero." While thirty people were injured, incredibly, no one was killed. Knowing that our live coverage of this event may have played a small role in saving lives was extremely gratifying.

MAY 15, 2013, GRANBURY/ LAKE PAT CLEBURNE TORNADOES

Some days look like textbook setups for severe weather, while other days are not so obvious, usually lacking at least one ingredient. With weak mid-level wind flow around 16,000', May 15, 2013, presented a case that didn't seem to portend tornadoes. Like a seasoned poker player, however, it revealed its hand late in the game. Meteorologist Jennifer Myers and I were on the air for this entire outbreak, beginning with late afternoon coverage during *FOX 4 News at Five O'Clock* and continuing well into the night. Initially, North Texas was under a severe thunderstorm watch in the afternoon and was upgraded to a tornado watch shortly before six o'clock in the evening, an hour before the stronger tornadoes started to develop. By six o'clock, super-cells were intensifying along a dry line northwest of the Metroplex, spawning the first tornado in Montague County. Soon after, a large cone tornado was spotted in Parker County between Millsap and Weatherford. Between six and seven o'clock, things escalated quickly, with additional supercells exploding along the dry line as far south as Central Texas. The atmosphere was unstable, but strong low-level wind shear was what tipped the scales toward tornadoes. Winds at the surface had shifted to the southeast by late afternoon and had increased to fifteen to twenty miles an hour, while up at 5,000' winds were blowing from the southwest at thirty-five to forty miles an hour. This clockwise "turning of the winds" with altitude produced a favorable environment for strongly rotating updrafts.

Between 6:30 and 7:30 p.m., at least four supercells capable of producing large hail, damaging winds, and tornadoes formed. To determine the threat level, we methodically tracked each cell on radar, plotted storm tracks with estimated arrival times, and relayed reports from storm chasers and law enforcement. By 7:45 p.m., we had three dangerous storms: one in Wise County, another in southern Parker County, and a third in Hood County, all likely producing tornadoes. Doppler radar was showing a "debris ball" or "tornado debris signature" with the Parker County storm, which meant we almost certainly had a tornado lofting debris into the air. (This will be discussed in greater detail in the next chapter.) Meanwhile, storm spotters were reporting a rotating

May 15, 2013 Live Coverage *(KDFW-FOX 4 News)*

wall cloud west of Granbury in Hood County, and radar displayed a strong rotational signature that had us very concerned. By eight o'clock in the evening, spotters confirmed a tornado northeast of Decatur in Wise County, moving east-southeast. To say we had our hands full is an understatement. Making matters worse, sunset was approaching, and darkness would make seeing the tornadoes difficult for spotters and those in the tornadoes' path.

"I WATCHED MY SON GET SUCKED OUT OF THE CLOSET"

Christy Green and her two sons, Dillon and Brandon, were at their home in the Rancho Brazos neighborhood just east of Granbury. At first, Christy thought it was an ordinary thunderstorm, but then Brandon's girlfriend texted to let him know there was a tornado on the opposite side of town. Christy took shelter in the closet as a precaution while her son, Dillon, a soon-to-be meteorology student at Texas A&M, and his older brother, Brandon, monitored the weather. Roughly ten minutes later, the tornado was heading their way, and the brothers fled to the closet. Within a few minutes, Christy says, they heard the tornado: "It sounded like roaring thunder that got louder and louder and never stopped."

She told the boys she loved them because she didn't know if they would survive the tornado. Within seconds, glass was shattering, the house was shaking, and the roof was about ready to go. At this point, the closet door started flapping open. "I told my oldest son, Brandon, to grab it and hold it. He did the best he could but before long it sucked him out. I watched him get sucked out of the closet. All I could see were his feet in the air." Seconds later, she and Dillon were airborne. "I was flying through the air like Dorothy and Toto in the *Wizard of Oz*. Things are just flying around. It was pitch black. I kept getting hit in the head by wood and other things. I told God I couldn't take it anymore and to just take me." Christy was then thrown into the ground feet first, her legs buried in debris to her knees.

"When it was over, it looked like a war zone. It looked like somebody had set off a bomb. Smoke was coming up from out of the ground. The house was gone. Nothing left but a slab." Dillon had landed across the street, while Brandon ended up in a field behind their home.

Amazingly, Christy was the only one of the three who sustained significant injuries; scans revealed a minor brain bleed and a fractured neck. She was also treated for a

Granbury Tornado *(Pikachu)*

gash on her forehead that needed several stitches and a knee laceration that required staples. After a few days in the hospital, she was discharged and reunited with her sons.

Home Swept Away in Rancho Brazos *(National Weather Service)*

The Rancho Brazos neighborhood was the epicenter of the tornado's wrath. Nearly a hundred homes, many built by Habitat for Humanity, were severely damaged or swept away by its 180-mile-an-hour winds. The EF-4 tornado claimed the lives of six people in the community, including an older couple who lived in a mobile home and another man who was caught outside in the tornado before he could get to shelter. The same supercell that generated the Granbury tornado later spun up an EF-3 tornado around 9:30 p.m. as it moved east into Johnson County, damaging homes and flipping several vehicles near Lake Pat Cleburne. The now mile-wide tornado then did something unusual. It abruptly turned north, putting itself on a collision course with the city of Cleburne. Fortunately, this enormous tornado fizzled out before reaching the more populated areas of Cleburne. In total, sixteen tornadoes hit North Texas, with the last one, an EF-1, touching down in Ennis shortly after midnight.

Christy built a new home on her lot, complete with an underground storm shelter. While the family's physical wounds healed quickly, the mental anguish lingered for years, particularly for Brandon. "He had anxiety and panic attacks. He ended up in the hospital several times after the tornado. Every time a storm would blow in, he'd suffer those panic attacks before they finally put him on medication." Christy claims the anxiety has finally dissipated but that all of them share a very healthy respect for the power of storms.

APRIL 29, 2017, EAST TEXAS OUTBREAK

A slow-moving cold front colliding with an explosive air mass over East Texas set the stage for an outbreak of tornadoes, including an intense EF-4 tornado that bulldozed its way from Eustace to West Canton and an EF-3 tornado that developed south of Canton and remained on the ground for eighty terrifying minutes. This outbreak, uniquely, was confined to a narrow corridor as tornadoes tracked repeatedly northward, parallel to State Highway 19 from northern Henderson County through Van Zandt and Rains counties.

April 29, 2017, seemed primed for severe weather from the get-go, as a plume of Gulf moisture pumped into North Texas. With dew points near 75°F, the tropical air condensed quickly upon being lifted, supporting low cloud bases. Tornadoes forming low to the ground would be difficult to spot in the heavily forested East Texas countryside. By the afternoon, surface winds had shifted to the southeast and an outflow boundary had set up along the I-20 corridor. The wind shift and low-level boundary favored significant low-level wind shear and were contributing factors for tornado development.

The first tornado of the day was an EF-0 that briefly touched down just south of Grand Saline. Another hour would pass before a second tornado was generated, another brief EF-0, just north of Canton. The third tornado was a far cry from the first two. It formed near Eustace in northern Henderson County around 5:30 p.m. and grew into an EF-4. This twister was a mile-wide beast with up to 180-mile-an-hour winds. Just twelve minutes later, a fourth tornado formed a few miles east of Eustace and tracked nearly parallel to the third tornado as it moved north into Van Zandt County.

TERRIFYING TALE FOR TWIN SISTERS

Laura Williams and her twin sister, Lisa Smith, lived roughly a hundred yards from each other on separate five-acre homesteads just off FM 1651, about seven miles south of Canton. As the first tornadoes hit the area, Laura received several conflicting text messages: "I was hearing about tornadoes north of town and west of town. Now we're hearing about another tornado. It was so confusing. I just thought people were getting their stories mixed up." Around 5:30 p.m., Laura's house lost electricity and Laura grew more concerned about the weather and her sister, who was home alone. She called Lisa to tell her to hide in her closet, and she headed to her own bedroom closet.

Around this time, Laura's husband, Biff, had just finished mowing the lawn and was on the porch scanning the sky for any signs of threatening weather. Shortly before six o'clock, Biff saw the sky turn pitch black. The tornado swallowed up the entire horizon as it tore utility poles out of the ground along FM 1651, just 150 yards from their home. "All of a sudden, my husband runs into the closet and shuts the door. When he closed the door, that's when the house started ripping apart."

Laura says she got down as low as she could, with her husband shielding her, and they prayed to God to help them. "We hear this growl. The scariest, scariest sound just engulfing the house. All I kept thinking about was my kids (who are grown and living on their own). Literally every second that would pass, I was amazed that we were still alive."

When it was over, everything was gone. The front half of the house, including the tile, carpet and hardwood floors, were wiped clean to the concrete slab. The rear portion of the home, where they had taken shelter, was now a pile of rubble. Incredibly, Laura and Biff had survived without any injuries. Seconds after the couple stepped out of their closet, Lisa emerged from beneath a pile of debris, lifting her closet door off her back. As Lisa rose to her feet, she yelled to Laura and Biff with her arms raised overhead to let them know she was alright. She, too, had miraculously survived.

As the violent EF-4 tornado was lifting, a destructive EF-3 wedge tornado packing winds up to 145 miles an hour developed twenty-five miles south of Canton at 6:08 p.m. This tornado churned

Tornado near Canton *(Pecos Hank)*

continuously for forty miles over the course of its eighty-minute lifespan, fortunately steering just east of the center of Canton. The twister produced some extremely tense moments for twenty people inside the Rustic Barn, who had been preparing for the Edgewood High School prom when the tornado hit. Everything except for the bathroom and closet they took shelter in was destroyed. The tornado continued to grow in size and intensity, easily ripping down massive transmission towers. It was nearly a mile wide when it reached I-20. One woman was killed when her car was thrown off the interstate into a field. A Dodge dealership was also destroyed, and some of the cars from its lot were tossed a half mile away. Professional storm chaser and filmmaker Hank Schyma (Pecos Hank) witnessed five tornadoes in Van Zandt County that day. He says it was one of the scariest chase days of his life due to the thick stands of trees that lined the roads and highways, at times completely obscuring his line of sight. Judging the conditions too dangerous to continue, Hank ended his chase in Fruitvale, just as darkness set in.

The twister hit the small town north of Canton head-on. Chasers often become first responders, and Hank immediately spotted a woman who needed his help. She had just crawled out from a pile of rubble that used to be her home. While she bled profusely from a gash in her head, she watched over her husband, who had broken his neck. As an ambulance transported her husband to a local hospital, Hank took the woman to receive medical treatment. Weeks later, Hank reunited with the couple over dinner and would later help raise funds for their medical bills.

While Hank drove to the hospital, Lisa and Laura set out in search of their parents. Traversing a landscape strewn with live power lines and downed trees, the twins feared the worst. An hour later, however, wandering down FM 1651, they found their parents, who had been searching for Lisa and Laura. Their eyes met and tears flowed. "My dad went to his knees as my mom cried out, 'My God, they're alive,'" says Laura. They had all survived and that was the only thing that mattered. Several days later, Laura got another unexpected surprise by way of a message on social media. A friend had seen photos of Laura and her family on Facebook. The person who posted the photos had found them in their yard, a hundred miles away in Oklahoma. Laura made contact with the poster and was able to recover some cherished photos she had thought were gone forever.

The sisters have since moved out of the Canton area but are eternally grateful for the outpouring of love and support from the entire community. Sadly, the East Texas outbreak resulted in four fatalities and more than fifty injuries. A total of seven tornadoes touched down between 4:15 and

7:15 p.m., including five in Van Zandt County. Two additional tornadoes, in Smith and Wood counties, touched down between 12:45 and 1:00 a.m.

OCTOBER 20, 2019, TORNADO OUTBREAK

On the evening of Friday, October 18, the outlook for severe weather on Sunday night was not overly concerning. The Storm Prediction Center (spc) had placed North Texas under a "Marginal Risk" for severe weather and the late-evening arrival was expected to sap some of the energy. As the weekend wore on, though, it became apparent that the atmosphere would be primed for severe weather. Strengthening jet stream winds, a surge of Gulf moisture, and high instability all pointed to the potential for potent storms. As a result, the spc upgraded the Dallas-Fort Worth area to a "Slight Risk," with an "Enhanced Risk" for Denton and Collin counties north into Oklahoma and Arkansas, beginning Sunday afternoon through late Sunday night. Clear skies and warm temperatures on Saturday and much of Sunday did not portend severe weather for Sunday night in the minds of most North Texans. Tens of thousands of people flocked to Fair Park to enjoy the closing weekend of the State Fair of Texas and to North Fort Worth to see the Blue Angels in flight at the Alliance Air Show. The perfect outdoor weekend culminated with a Sunday night marquee matchup between the Cowboys and Eagles at AT&T Stadium. The game, which drew 1.5 million viewers in the Metroplex, may have been a blessing and a curse. While it likely kept many people at home glued to their TV sets, it also diverted attention away from the storms that were beginning to blossom southwest of the Metroplex just before sunset.

Fifteen minutes prior to kickoff at AT&T Stadium, a tornado watch was issued for North Texas that highlighted the potential of damaging wind gusts up to eighty miles an hour, large hail, and a tornado threat that would increase into late evening. At that time, severe thunderstorms were rumbling just north and west of the Metroplex, with gusty winds upwards of fifty miles an hour and reports of up to golf-ball-sized hail. Between 7:30

and 8:30 p.m., severe thunderstorms rolled eastward into Denton, Tarrant, and Johnson counties, prompting warnings for wind gusts up to sixty miles an hour and hail up to the size of half dollars. The situation escalated quickly around nine o'clock, as a pair of supercells moved from Tarrant and Johnson counties east into Dallas and Ellis counties. Radar revealed that rotation in the storms was increasing as they encountered a very moist, unstable environment with impressive low-level wind shear. Meteorologist Ali Turiano and I were focused on these storms as we prepared for *FOX 4 News at Nine O'Clock*. Growing increasingly concerned with the strong rotation observed in Northwest Dallas, the National Weather Service pulled the trigger and issued a tornado warning for much of Dallas County north of downtown Dallas. We led our newscast with the warning, zooming in on the ominous storm, where we pinpointed the tornadic circulation near I-35E and Walnut Hill Lane and urged people to take cover. Within a couple minutes, we received reports of a confirmed tornado with debris just northwest of Love Field. We immediately understood that this was an extremely dangerous situation. The tornado was going to track across one of the most populated areas in Texas under the cover of darkness, and it was advancing at a brisk thirty-mile-an-hour pace, covering a half mile each minute. We followed the tornado as it crossed Harry Hines, the Tollway, and Central Expressway. I advised viewers to take cover in their shelters and to call or text family and friends watching the game, who might be oblivious to the danger they were in.

"IT SOUNDED LIKE A JET WAS TAKING OFF"

Blake Wiley was watching the Cowboys game that evening at his North Dallas home, just south of Royal Lane, near St. Mark's School of Texas. He said emergency alerts on his phone and texts his wife was receiving from neighbors prompted them to change the channel in search of information. "That's when I heard the tornado was going to hit between Walnut Hill and Forest and I knew that it was very close." Blake scrambled to wake up his children, ages five and three, and joined his wife, Eve, and his mother as they all huddled into a small powder room under the staircase. Within a few minutes, everything went dark. "All the lights went out. Then it got eerily quiet. That's when I thought, this is about to hit us. Then [I heard] a loud whistling sound and I started praying. Our ears popped like when you're landing in an airplane. That must have been when we lost our roof and we could hear the wind inside the house. It was like a jet was taking off outside the door. The wind was coming through the bottom of the bathroom door. It lasted about twelve seconds, and then it was gone."

When Blake opened the door, he immediately detected smoke in the air. Fearing the house might be on fire, he surveyed the upstairs and downstairs, and determined it was just dust and debris suspended in the air. When he looked out the front window, he saw a car resting on its side in his front yard. "I saw the lights were blinking and the airbags were deployed. At first, I didn't know that it was my mom's car. I went outside and looked in it and no one was inside." Within five minutes, firefighters were on the scene. They promptly ordered the Wileys out of their house after detecting gas leaking from the car that was now just feet from the front door.

The Wiley Home

Sunrise on Monday morning revealed the true extent of the damage to the Wiley home and hundreds of others in the Preston Hollow neighborhood. The Dallas tornado had carved a path of destruction three-quarters of a mile wide and over fifteen miles long, from Northwest Dallas to Richardson. Rated an EF-3 with estimated winds up to 140 miles an hour, the twister damaged or destroyed more than nine hundred structures; it blew out windows, overturned vehicles, sheared off trees, downed power lines, and ripped roofs off of homes, businesses, churches, schools, and apartment buildings.

North Dallas Tornado Damage *(Andy Luten)*

The initial tornado damage in Dallas was observed near I-35E and Walnut Hill. The tornado quickly gained strength as it moved across Harry Hines Boulevard. From there, it plowed through neighborhoods and several schools between Walnut Hill and Royal Lane, steamrolled east through Preston Hollow, then veered to the northeast, where it ripped a gaping hole in the Home Depot near US 75 and Forest Lane. After crossing US 75, it passed just southeast of the High Five Interchange, inflicting heavy damage on several homes near Audelia and Buckingham in Richardson before finally roping out. Over the next three days, National Weather Service survey crews confirmed nine more tornadoes. Four EF-0 spin-ups hit Allen, Ferris, rural Kaufman County, and Wills Point; four EF-1 twisters touched down in Midlothian, Rowlett, Rockwall, and rural Kaufman County, and a high-end EF-2 with 135-mile-an-hour winds traveled two-and-a-half miles through Garland.

The Insurance Council of Texas estimated that losses from the October 20, 2019, tornado swarm were at least $2 billion, making it the costliest tornado outbreak in Texas history. Despite the trail of destruction and thousands of people who were in the crosshairs of these fast-moving tornadoes, there were no life-threatening injuries or fatalities. The fact that the violent storms struck on a Sunday night while many residents were at home certainly played a crucial factor in this. The Dallas tornado crossed several major highways and thoroughfares during its fifteen-mile rampage, including Interstate 35E, Harry Hines Boulevard, Preston Road, Royal Lane, the Dallas North Tollway, Central Expressway, and Interstate 635. A tornado of this magnitude barreling across commuter-clogged roads during the weekday rush hour would have likely resulted in mass casualties.

TO INTERRUPT OR NOT TO INTERRUPT

This tornado outbreak highlighted the challenges local TV stations face when severe weather hits during primetime programming. In this case, NBC was airing the Cowboys–Eagles game while dangerous storms blazed through densely populated areas. Aside from catching a couple abbreviated weather updates, many viewers watching the football game were left in an information void. One of those viewers was Nancy Fluce, who was at her Preston Hollow home watching the game while talking with her sister on the phone. "The sirens were going off and I had looked at the TV and they reported a possible tornado at Love Field." That was the only update she recalls seeing during the game. When the sirens went silent, she figured they were not in the tornado's path. "Then it got eerily still, and the house alarm sounded, and the dogs started barking. I could feel the pressure. Within seconds, the wind became extreme and I could hear glass cracking and shattering." Nancy, her husband, and their dogs took shelter in a closet beneath the stairs just as the chimney collapsed in the dining room at the front of the house. Nancy's son had barely made it back to his garage apartment behind their home when the tornado hit. She feels lucky and blessed that they all escaped without a scratch.

Collapsed Chimney in the Fluce Home

Timing was on our side on this night. FOX 4's regularly scheduled newscast at nine coincided with the first tornado warning issued for Dallas County, and we continued with five hours of live, nearly continuous weather coverage. Countless times over the years we have taken a lot of heat for interrupting sporting events and highly rated television shows to provide severe-weather updates. We've also been chastised for not breaking in often enough. The reality is that while storms may be raging in one or two counties, the weather is typically tranquil over the remaining 95% of North Texas. When you decide to break into programming, you have to interrupt for everyone. Some folks are understanding, but there is always a vocal contingent that expresses its extreme displeasure with angry and often profane emails, phone calls, and social media posts directed at those delivering the message, the local meteorologists.

The decision of whether to cut into programming is one we take seriously. The meteorologist on duty will always confer with at least one manager and the news director. Every situation is different, and we carefully weigh the options. If, for instance, a severe thunderstorm with sixty-mile-an-hour wind gusts does not pose a significant threat to life and property, we may choose to stand down and keep viewers informed with live updates on our various social media platforms. Tornado warnings are a different story, and they are given special attention. Many years ago, I interrupted a highly anticipated episode of *Glee* when a tornado warning was issued for Tarrant County. It was after dark, the storm was moving east at forty miles an hour, and radar was indicating strong rotation. While the tornado was not confirmed, trained spotters were reporting debris. Fortunately, the tornado never materialized, and no significant damage was reported. Unfortunately for our station, the prolonged severe weather interruption sparked an angry backlash from thousands of disgruntled viewers. Seconds matter for those in the path of these potentially deadly storms. When lives are at stake, we can't afford to wait for confirmation. The events of October 20, 2019, have kindled several discussions between management and our weather staff about interrupting programming for emergency weather coverage. FOX 4 understands how important programming is to its viewers and makes every attempt to minimize disruptions. However, at the end of the day, human life takes precedence over programming, and we will interrupt if we feel it's prudent to do so, to give our viewers up-to-the-minute information that will help them stay safe.

CHAPTER 6

EYEING THE STORM

TORNADO DETECTION AND CHASING

On March 20, 1948, a powerful tornado hit Tinker Air Force Base in Oklahoma City without warning, tossing military aircraft like toys, inflicting numerous injuries, and causing $10 million in damage. Over the next few days, Air Force Captain Robert Miller and Major Ernest Fawbush, who had both been on duty when the tornado hit, feverishly studied surface and upper-air weather charts and examined data from past tornadoes, determined to learn more about what had happened. On March 25, Miller and Fawbush recognized a similar threatening weather pattern taking shape and did something that had never been done before—they issued a tornado forecast. Up until 1948, the Weather Bureau (now called the National Weather Service) had banned forecasters from even using the word "tornado" for fear that it would cause mass panic. Sensing that another tornado was possible, the Air Force meteorologists were given the green light by Commanding General Frederick Borum to issue the nation's first tornado forecast.

Incredibly, just after six o'clock in the evening, a strong tornado struck the base again, the second time in five days. Because of Miller and Fawbush's bold forecast, preparations had been made to secure aircraft, move personnel to safe locations, and divert inbound flights. While the tornado resulted in $6 million worth of damage, there were no injuries, thanks to the carefully worded warning. The pioneering achievement proved that severe weather could be anticipated with a reasonable degree of accuracy and focused national attention on forecasting impending tornadoes and warning the public of their inherent danger.

THE EVOLUTION OF WEATHER RADAR

The most valuable tool in a meteorologist's arsenal for pinpointing and tracking severe weather is radar. Meteorologists stumbled upon its practical use for detecting precipitation during World War II. Navy meteorologists, while using radar to detect enemy warships and aircraft, would often notice strange echoes appearing on their displays. After thorough investigation, they discovered the extraneous echoes that were cropping up were areas of precipitation. After the war, in 1946, the Weather Bureau obtained twenty-five AN/APS-2F radars from the Navy. These aircraft radars were modified for meteorological use and deployed across the country as weather surveillance radars (WSRs).

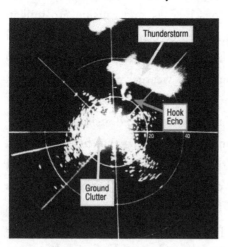

Illinois State Water Survey *(University of Illinois)*

One of the first significant radar revelations occurred on April 9, 1953, when an electronics technician observed something fascinating on an Illinois State Water Survey radar located at Willard Airport, south of Champaign, Illinois. Don Staggs, the radar technician, was working late to perform repairs when he noticed an interesting echo on the scope as a major tornado passed twenty-five miles north of the radar site. Staggs recorded the strange, hook-like appendage on the thunderstorm cell using the mounted thirty-five millimeter camera. One month later, two more hook-shaped echoes were observed and photographed during large tornado outbreaks in Waco, Texas, and Worcester, Massachusetts, proving that radar could be helpful in detecting tornadoes. (Not all hook echoes indicate a tornado and not all tornadoes will display a hook echo on radar.)

The violent F5 tornado that hit Waco killed 114 people and triggered a public outcry for the state of Texas to get serious about storm readiness and develop public warning procedures. This resulted in the creation of the Texas Tornado Radar Warning Network, the first of its kind in the United States. The program enabled cities to buy military surplus radars for the price of installation and modification, and it incorporated volunteer storm spotters.

On April 5, 1956, a watershed moment took place in College Station, Texas. Meteorologists operating the Texas A&M University radar observed tall, hook-shaped echoes and promptly warned the Bryan Police Department and the local school district that a tornado would likely occur within thirty minutes. Roughly twenty-five minutes later, a tornado touched down, inflicting $250,000 worth of damage on Bryan and College Station but causing no injuries. This is considered the first warning based solely on the interpretation of radar data.

While the 1953 tornado tragedies of Waco, Texas; Worcester, Massachusetts; and Flint, Michigan, highlighted the need to create a national radar network, a series of hurricanes that struck the East Coast in 1954 and 1955 finally prompted Congress to act. Hurricanes Carol and Edna struck the New England coast within eleven days of each other in late summer 1954. Hazel, a powerful category 4 storm, slammed into the North Carolina coast that October. In 1955, the onslaught continued, with hurricanes Connie and Diane making back-to-back landfalls, causing major flooding in the Northeast and resulting in $1 billion in damages and 258 fatalities.

In 1956, Congress approved funds for the design and installation of the Weather Bureau's first flagship radar, the WSR-57 (Weather Surveillance Radar—1957), as well as for staff to operate it. It was crude compared to today's modern radar technology. Storms, seen as green blotches, were tracked across the radar screen using grease pencils, and forecasters had to manually turn a crank to adjust the radar's scan elevation. The first WSR-57 became operational in Miami, in June of 1959, and was followed by the installation of thirty-one more radars throughout the early 1960s. Just as the Weather Bureau was establishing the nation's first radar network, NASA launched the Television Infrared Observation Satellite (TIROS-I), the world's first successful weather satellite. Although TIROS-I operated for only seventy-eight days, it sent back 19,389 satellite photos, ushering in the era of monitoring weather from space.

The WSR-57 served as the workhorse radar for the National Weather Service for more than thirty-five years. In the 1990s, the National Weather Service unveiled its replacement, the NEXRAD (Next-Generation Radar) system, a network of 159 high-resolution Doppler radars that were deployed across the country. NEXRAD proved to be a major upgrade over the World War II technology, providing much greater resolution and sensitivity and allowing radar operators to see features like gust fronts and cold fronts, which had not previously been visible on radar. Perhaps the most important improvement over conventional radar was that NEXRAD provided Doppler velocity data, which would enable operators to detect rotation in the storm at different elevations and improve tornado prediction.

THE DOPPLER EFFECT

Doppler radar systems utilize the "Doppler effect" to provide information regarding the motion of precipitation along the radar beam. You experience the Doppler effect when a train or emergency vehicle approaches and then passes you. Consider a fire engine, with sirens blaring, driving toward you. The pitch of the siren is higher as it moves toward you because of the compression of the sound waves. As the fire engine moves away from you, the sound waves are stretched, resulting in a lower pitch. The same effect takes place in the atmosphere as a radar pulse strikes an object, such as a raindrop, and is reflected back toward the radar dish. The frequency shift between the transmitted radio wave and the reflected echo the radar receives is directly related to the speed and motion of the precipitation. If the raindrops are moving toward the radar, the frequency of the reflected echo will be higher. On the other hand, if the raindrops are moving away from the radar, the frequency of the reflected echo will be lower. This ability to detect the wind's orientation enables meteorologists to detect rotation within a storm that can lead to tornado formation.

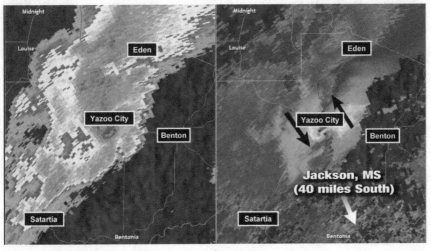

National Weather Service (*Jackson, MS*)

The side-by-side images above were taken from the Brandon, Mississippi, radar on April 24, 2010. At the time, a nearly two-mile-wide EF-4 tornado was roaring through Yazoo City, Mississippi. The standard reflectivity image on the left, which most viewers are familiar with, doesn't display the classic hook echo. The HP supercell was producing heavy curtains of rain that were circulating around the massive twister. In the image on the right, the bright couplet at the center indicates very strong inbound and outbound velocities (the arrows indicate motion toward the radar and motion away from the radar). The ability to detect strong rotation in this storm enabled forecasters to issue timely tornado warnings that, considering this destructive tornado stayed on the ground for 149 miles, likely saved many lives.

Radar Receiving Dish Inside Radome *(NWS)*

Today's NEXRAD network also employs dual-polarization technology that transmits and receives both horizontally and vertically oriented pulses. This provides measurements of the horizontal and vertical dimensions of targets to help meteorologists distinguish between heavy rain and hail, and sleet and snow. This "dual-pol" technology also provides improved detection of non-meteorological echoes like insects, birds, and debris associated with tornadoes. This is extremely valuable for tornado detection. The technical term for this dual-pol radar technology is "correlation coefficient." Correlation coefficient is a measure of how uniform the features being observed by radar are relative to one another. High correlation coefficients indicate pure rain or pure snow because of the similarity of raindrops to other raindrops and snowflakes to other snowflakes. On the flip side, the irregular size, shape, and tumbling nature of tornado debris produces low correlation coefficients.

The 3-D correlation coefficient radar image from the Dallas tornado on October 20, 2019, clearly depicts tornado debris (tree branches, shingles, and other building materials) with bright pixels (circled) in the lower foreground amidst a large area of precipitation in the more uniformly shaded background. The tornado debris was lofted 20,000' into the air, a testament to the lifting forces of this tornado, which easily peeled roofs off of hundreds of well-built homes. A meteorologist with multiple radar products is like a detective with multiple pieces of evidence; both are able to make a stronger case. If a radar operator sees an area of high reflectivity colocated with an area of strong rotation and low correlation coefficient, they know a tornado is likely lofting debris into the air.

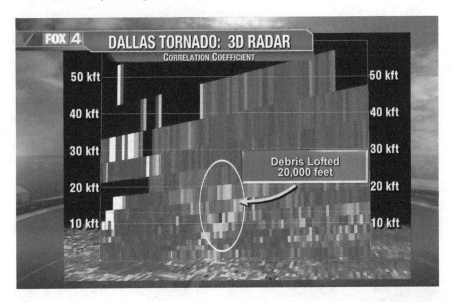

NORTH TEXAS GETS MORE "EYES IN THE SKY"

Even the most sophisticated radar technology has limitations. First, the farther away a storm is from the radar, the more difficult it is to see many of the important storm features. A radar beam is similar to the light emitted by a flashlight, spreading out as it moves farther from the source. The radar beam expands roughly 1,000' for every ten miles from the radar site, so the initially concentrated beam becomes diffused at long distances, causing small features to slip out of focus. Second, the radar beam travels in a straight line and the earth curves away beneath it. At a distance of fifty miles from the radar site, the radar beam is roughly 4,500' above the ground, and at a distance of one hundred miles, the radar beam is aiming 12,600' above the surface. At this distance, the radar is overshooting the critical lower and middle portions of the storm, where rotation that precedes tornado development cannot be seen.

Prior to 2012, much of North Texas was served by three weather radars. The main NEXRAD radar was located near Burleson, south of Fort Worth, and two FAA terminal Doppler radars were stationed at DFW Airport and Dallas Love Field. In 2011, a unique partnership was formed between the Center for Collaborative Adaptive Sensing of the Atmosphere (CASA),

the National Weather Service, and the North Central Texas Council of Governments (NCT-COG) to expand radar coverage in North Texas. The plan was to install a dense network of eight low-power, short-range, dual-polarization Doppler radars that would scan a much larger portion of North Texas. The first radar was deployed at the University of Texas at Arlington in 2012 and has been followed by the installation of six additional radars in the cities of Denton, Midlothian, Fort Worth, Cleburne, Mesquite, and Garland.

This radar network has already paid dividends for local meteorologists at the National Weather Service in Fort Worth by enabling

Radar Installation in Cleburne *(CASA)*

them to probe a greater portion of the lower atmosphere to detect low-level rotation associated with tornadoes. The CASA radar in Midlothian helped pinpoint the location and movement of a compact EF-0 tornado that tracked just south of the Metroplex in northeastern Johnson County on the evening of January 15, 2017. Knowing the precise location and movement of this tornado was of particular importance because one hundred thousand fans were dispersing from AT&T Stadium in Arlington after a playoff game between the Cowboys and Packers. The overlapping radar coverage also provides a much more detailed look at near-surface wind fields, allowing forecasters to assess the potential for damaging wind gusts and locate boundaries that may serve as triggers for storm development. These radars provide more accurate estimates of rainfall, which can lead to more advance warning in flash flood events.

THE NEED FOR BOOTS ON THE GROUND

Hollywood films tend to portray storm chasers as thrill-seekers and adrenaline junkies who throw caution to the wind as they barrel down country roads while dodging flying debris in order to capture the "money shot" of a tornado in all its glory. The vast majority of storm chasers strive to put their safety first. Over the years, we have been extremely fortunate to have several dedicated, experienced chasers, including Russ Contreras, Kevin Saunders, Gene Yates, Justin Cullum, Victor Florez, Nick Copeland, Nick Busby, Paul Stofer, Aaron Estman, Mark Drees, Chad Casey and Traci Tuttle who have logged thousands of miles on nearly every farm-to-market road from Waco to Wichita Falls. These volunteers put themselves in harm's way to get eyes on the storm and provide us with critical, up-to-the-minute reports. Many of them regularly attend chaser conferences and National Weather Service SKYWARN Spotter Training sessions to learn about the fundamentals of storm structure, how to identify potential severe-weather features, and how to report information to local NWS offices. SKYWARN training sessions are geared toward "volunteer spotters," citizens who intend not to chase but rather to report on weather conditions from their residence or place of business.

Storm chasers and spotters play a vital role in the process of warning the public of potentially dangerous weather. The only way to be 100% certain that a tornado is on the ground is to have eyes on it. A well-trained spotter can provide critical information in real time, such as hail size, wind speed and direction estimates, whether a wall cloud is rotating and if a condensation funnel has formed, and the size of the tornado and any damage it may have produced. When issuing warnings, NWS meteorologists can then use more descriptive language, such as "trained spotters report a large cone tornado with debris" to add veracity and prompt citizens to take swift, lifesaving actions.

THE INHERENT DANGER OF CHASING

Storm chasing carries an element of risk. Tornadic storms often display unpredictable motion, with sudden twists, turns, and acceleration. Maps and GPS may give you confidence that a road will lead you to safety, but storm chaser Hank Schyma, better known as Pecos Hank, advises that unforeseen hazards and escape route entrapments can arise. "[Maps and GPS] can't predict flooded roads, that a stalled one-hundred-coach train may be blocking your retreat, when a bridge is out, when fallen trees are blocking your path, or a random deep pothole…anxious to tear your wheel off its axle." Unfortunately, there are some who choose to tempt fate by punching through the angry belly of intense storms. "Core-punching" involves driving through the core of the heaviest rain and largest hail to get to a better storm-viewing location. This dangerous act may subject storm chasers to near-zero visibility in blinding sheets of rain and a pummeling by hail the size of baseballs. If they emerge safely on the other side of the storm, they may find themselves in the direct path of a tornado.

Most responsible chasers will not attempt stunts like this. Instead, they formulate a plan for the day, beginning by thoroughly analyzing the weather data to determine where storms may initiate and how things may unfold. On active days, they will likely chase with at least one other person in the vehicle, who will navigate and monitor the storms on radar, although even the best-laid plans can be foiled, as weather conditions can change suddenly and deteriorate rapidly. The combination of these elements and a highly anticipated severe-weather event that draws hundreds of trained chasers and wannabe chasers from all over the country can create epic log jams on two-lane country roads and be a recipe for disaster. Sadly, this happened on May 31, 2013, in El Reno, Oklahoma. Veteran storm chaser Tim Marshall describes the storm as wildly unpredictable: "Sometimes storms will grow much larger, turn in a different direction or accelerate. This one did all of that." The National Weather Service office in Norman, Oklahoma, produced an excellent video (scan the QR code below to watch) that chronicles the life cycle of the most dangerous tornado in storm-observing history and the lessons learned that deadly day in El Reno.

NWS Norman, OK
(El Reno Lessons Learned)

The El Reno tornado was an exceptionally large (2.6 miles wide), violent tornado with several intense sub-vortices moving erratically at blistering speeds around the main circulation. The main tornado was also moving in a highly unpredictable manner. It changed direction multiple times, looped at least once, and at times was nearly stationary and would then accelerate to speeds of up to fifty-five miles an hour. On top of this, a smaller but powerful 150-mile-an-hour anticyclonic (clockwise-spinning) tornado had formed five miles away. To make matters even more complicated, the tornado was rain wrapped at times and suddenly grew much larger with fierce winds extending well beyond the visible condensation funnel. Many storm chasers were caught completely off guard and found themselves in an impossible situation that forced them to abandon the chase and flee for their lives. Four chasers perished, including members of the TWISTEX research team led by legendary chaser Tim Samaras, along with his son Paul and his chase partner, Carl Young. Their deaths stunned and deeply saddened the entire storm-chasing community.

Tim Samaras was a seasoned, savvy chaser with twenty-five years of experience pursuing storms. He often trekked twenty-five thousand miles in a year for his research. Tim was a talented electrical engineer who helped build a model of the center wing tank to determine what had led to the crash of TWA Flight 800 and worked on classified technology to detect suicide bombers for the Department of Homeland Security. But Tim's true passion was chasing tornadoes in an effort to take groundbreaking measurements of near-surface winds and atmospheric pressure inside one of them. To do this, he built sophisticated probes that he called "turtles," which were six-inch-tall conical weather stations encased in steel that he would attempt to drop in the path of an oncoming tornado. For decades, teams of scientists had tried to do something similar but had failed. Tim, however, had an uncanny knack for being in the right place at the right time and made several successful deployments that resulted in direct hits to his probes.

TWISTEX Research Team *(Tony Laubach)*

Tim made history in 2003 when one of his probes scored a direct hit from an F4 tornado in Manchester, South Dakota, and was successful in measuring a 100 mb pressure drop. In a 2005 interview I conducted with Tim for FOX 4 News, he described the accomplishment: "We knew we had a direct hit and when we looked at the data and saw a 100 mb pressure drop, we were just flabbergasted. It's like stepping into an elevator and shooting yourself up 1,000' in ten seconds and then your ears pop. That's the kind of pressure change we are talking about." Tim, though driven and passionate about his work, was not known as a daredevil. In fact, there were many times he refused to deploy his probes because doing so would have been too dangerous. Video from storm chaser Dan Robinson's rear-facing dash camera captured the last images of Tim's car disappearing behind heavy curtains of windswept rain. The car's occupants likely could not escape the powerful storm inflow that was generating a sixty- to eighty-mile-an-hour headwind, particularly considering they were in a small four-cylinder Chevy Cobalt loaded with three passengers and heavy equipment. The TWISTEX crew, among others, probably misjudged the size of the tornadic wind field, which expanded rapidly and cut off their escape route. Josh Wurman, who operates a fleet of mobile radars for his Center for Severe Weather Research, was scanning the immense tornado. High-resolution data from his Doppler on wheels indicated that one of the satellite vortices, traveling at an incredible 176 miles an hour, struck Tim's vehicle and threw it nearly 2,000'. A larger Weather Channel chase vehicle was also overtaken by an intense sub-vortex, rolled, and flattened, but amazingly, all of its occupants survived.

Tim Marshall was only one mile northeast of Samaras and his group and was one of many chasers who were extremely fortunate to escape the massive, deadly twister. He says El Reno was a wake-up call for chasers: "Tornadoes are erratic. You have to anticipate and think ahead. There are certain situations you simply have to abort. Back out and live to chase another day."

TAKING RADAR TO THE STORM

Josh Wurman recalls El Reno being a "crazy day" with far too many chasers converging on a very dangerous storm. Josh sensed the vulnerable situation he would be in if he got too close, so he stayed well back. Probing the storm from two miles away, he captured phenomenal images of the supercell's complex wind field and the deadly tornado it unleashed. One of Josh's Dopplers on Wheels (DOWS), the Rapid DOW, simultaneously transmits six beams that can sweep the sky in six seconds. His DOW technology has come a long way since the first primitive mobile radar he built. "We used 486 PCs and CRT monitors and used duct tape and Velcro to hold things together," he says. A former scientist for the National Center for Atmospheric Research (NCAR), he went to the scrap heap and salvaged a variety of surplus parts from old equipment to build the first prototype DOW. In late spring 1995, Josh and his small crew rolled out the DOW for the first time, but the maiden voyage in Kansas ended abruptly when the radar dish went haywire. "We had a VHF radio. Anytime someone keyed the radio, it made the antenna spin and go crazy and it broke the radar because we hadn't grounded it right." After making repairs, the crew hit the road again and scored their first tornado in Dimmit, Texas, on June 2, 1995. Josh calls it one of the most rewarding and exciting days of his life: "We made the first three-dimensional maps of the winds and the debris

DOW Probes a Tornado *(Herb Stein/CSWR)*

cloud and the evolution of that storm." In the twenty-five years since that first tornado intercept, Josh and his fleet of DOWs have scanned more than two hundred tornadoes and made important revelations. "We've observed that tornado winds change pretty rapidly. The strongest winds are actually close to the ground. The DOW measurements showed they are down close to house height (below 30') and not aloft like a hurricane. The other thing we've learned is the average tornado is stronger than has been thought. The average tornado is capable of doing EF-2 or EF-3 damage."

BOLDLY GOING WHERE NO MAN CAN GO

Getting scientists up close to sample the atmosphere surrounding tornado-spawning supercells is very risky business, which is why meteorologists at the University of Nebraska (Lincoln) employ drone technology (unmanned aircraft) to gather critical data around the storm. While unmanned aircraft have been used before in tornado research, the TORUS project (Targeted Observation by Radar and Unmanned aircraft systems of Supercells) is using some of the most advanced drone technology to date, thanks to collaboration with Dr. Brian Argrow and his colleagues in the Aerospace Engineering Department at the University of Colorado Boulder. This third generation of drones can be launched directly from vehicles using a pneumatic launching system (a catapult mounted on the roof). These drones have also proven to be durable, flying through hail and seventy-six-mile-an-hour gusts of wind. (Scan the QR code below to watch one of these drones in flight as it approaches a supercell in South Dakota.) The project, funded by the National Science Foundation, and led by UNL's Dr. Adam Houston, involves more than fifty scientists and students from four universities. TORUS deploys a broad suite of cutting-edge instrumentation, including four drones, three mobile radars, eight mobile mesonets (trucks mounted with weather instrumentation), three mobile sounding systems (to release weather balloons), and the NOAA P3 manned aircraft. The researchers

Scan for Video
Drone in flight
(UNL/TORUS)

journeyed nine thousand miles from South Dakota to Texas during their five-week mission in 2019, intercepting twenty-one storms and capturing some of the most detailed near-storm measurements of supercells ever gathered. Houston believes drone technology will play a greater role in the next generation of weather and storm surveillance, and this project will help scientists advance their understanding of how to use these aircraft. TORUS conducted a second data-gathering mission in the spring of 2020.

THE NEED FOR BETTER WARNINGS

Advances in radar technology and a deeper understanding of how severe thunderstorms produce tornadoes have unquestionably enabled forecasters to issue timelier warnings to the public. During the Super Outbreak of 1974, in which 148 tornadoes were spawned within twenty-four hours, the average alert gave people only a few minutes of advance notice. Today, the average warning lead time is fourteen minutes. That's the good news. The bad news is that the false-alarm rate is 70% to 75%, meaning that for nearly three out of every four tornado warnings issued, a tornado does not occur. Prominent researchers like Josh Wurman and Paul Markowski see a major additional flaw in the warning system: too many people are being included in the warning polygons (the area under a warning is outlined by a polygon). According to Markowski, this has to do with how the National Weather Service grades its forecasters on warnings issued. "[Forecasters] are issuing huge polygons. All they need to have is one hit in that giant net that they've cast, and it's graded a good warning. That's not the best public service." Markowski believes that if a storm has strong rotation and a warning is issued, the warning should be considered good, regardless of whether or not the storm produces a tornado. Josh Wurman recalls the EF-5 tornado that hit Moore, Oklahoma, in 2013 and generated mass panic because so many people were placed under an official warning: "All of Oklahoma City was in this tornado warning and people were told there was this big EF-5 tornado coming toward them. People fled their homes and there was gridlock on the highways. If that warning could have been trimmed down, then you wouldn't have had the whole city evacuating." The need to make improvements to the weather warning process presents a dilemma. On one hand, forecasters want to increase lead time, especially for hospitals, assisted-living facilities, and large venues like stadiums and amusement parks that require significantly more time to get people to safe shelter. On the other hand, there's a call to issue smaller-scale polygons that warn fewer people and lower the false-alarm rate.

THE NEW "WARN ON FORECAST" APPROACH

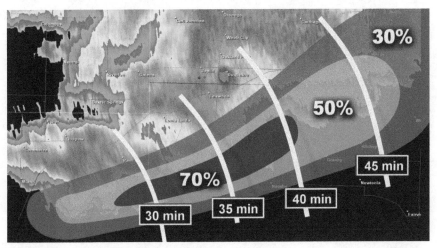

Warn on Forecast Conceptual Model

The National Severe Storms Laboratory (NSSL) in Norman, Oklahoma, is home to some of the brightest minds in severe-weather research. These scientists recognize the dilemma and are working toward a solution based on a new approach called "Warn on Forecast." Presently, tornado warnings are issued based on a "warn on detection" method, whereby if Doppler radar indicates strong, persistent rotation or a tornado is spotted, a warning is issued. But NSSL researchers believe this warning strategy is reaching a plateau and that further increases in effective lead time will be difficult to achieve. Their new "Warn on Forecast" philosophy will use ultra-high-resolution computer models that predict specific weather hazards, such as tornadoes and large hail, thirty to sixty minutes before they form. These advanced computer models will predict the probability of a hazard occurring and the confidence in the path of the storm, and will continually make adjustments to the forecast based on new weather observations and real-time radar scans of the storm.

The figure above represents a conceptual model that predicts the path of a potentially tornadic supercell. The new Warn on Forecast tornado projections will look more like hurricane forecasts. Forecasters will create a cone marking the potential path of the tornado, and additional contours within the cone will be labeled with percentages to define the chances of the tornado hitting those specific areas. This will reduce the number of people affected by the warning and help minimize false alarms. The new warning system, which may become operational as early as 2023, has the potential to be a game changer in delivering lifesaving weather information to those in peril.

CHAPTER 7

MEGA HAIL AND DOWNBURST WINDS

I n 1974, Nolan Ryan became the first Major League pitcher to break the one-hundred-mile-an-hour barrier when one of his fastballs was clocked at 100.8 miles per hour. At that speed, the ball takes a mere 0.41 seconds to cross home plate after leaving the pitcher's hand.

Mother Nature also throws a pretty impressive fastball. A smooth, dense hailstone the size of a baseball that doesn't encounter a great deal of wind resistance hits the surface with a speed of approximately eighty-five miles an hour. If the hailstones grow to the size of softballs with a diameter of roughly 4", they can attain speeds over one hundred miles an hour. The impact felt by someone unfortunate enough to be caught in the hailstorm would be equivalent to getting beaned by the "Ryan Express." Large hail falling at those velocities can be very destructive. If you live in North Texas, you have already experienced or likely will experience this firsthand. The Storm Prediction Center map below clearly shows that North Texas lies squarely in an active zone for hail. Based on all the reports of severe hail (1" or larger) in North Texas from 1986 to 2015, there was an average of four to five days of quarter-sized or larger hail within twenty-five miles of any point in our region. In other words, if you were to pick a point and go twenty-five miles in all directions, you would have a circular area that is roughly the size of Dallas and Tarrant counties combined. Interestingly, "giant hail" events, those that produce hail the size of tennis balls, baseballs, or softballs, are rare. They occur less frequently on average than tornadoes. In the same thirty-year period mentioned above, North Texas averaged only 0.6 to 0.8 days per year for hail measuring more than 2" in diameter, while it averaged one day per year for tornadoes. "Severe" wind events, those that produce wind gusts of fifty-seven miles an hour or greater, trump both hail and tornadoes. On average, severe thunderstorms produce five to six days of damaging wind gusts each year.

THE HIGH COSTS OF HAIL DAMAGE

According to Verisk, a company that monitors hail reports throughout the United States, Texas far exceeds every other state in the country in property damage due to hail. For the years 2000 through 2013, Texas averaged $859 million in insured property losses annually due to hail. (Minnesota was a distant second at $252 million.) Losses in recent years have dwarfed this number. In 2016, Texas was hit by some extreme hailstorms, including a $1.36 billion storm that pounded the San Antonio area. According to the Insurance Council of Texas, the Lone Star State recorded a whopping $5 billion in insured hail losses and another $1 billion in losses due to wind claims in 2016 alone. Even those who have not suffered wind or hail damage to their homes have been hit hard in their pocketbook; the historic losses are being blamed for skyrocketing insurance premiums that often come with high deductibles. According to 2018 figures from Insurance.com, Texans pay the seventh-highest homeowners' insurance rates in the country, dishing out an average of $1,945 annually for a $200,000 dwelling. Hurricane-prone Florida homeowners claim the nation's highest rates at $3,575 annually.

SIMULATING FALLING HAIL IN THE LABORATORY

We are frequently asked about the one-inch-diameter threshold for defining "severe hail." While the threshold seems arbitrary, research has revealed that "significant damage" generally does not occur from hail smaller than an inch. Hailstones the size of quarters or larger can inflict damage to cars, homes, and crops. Haag Engineering in Dallas has conducted hail damage research for decades, including testing roofing products to determine their susceptibility to hail. Using a mechanical device called an ice ball launcher, they fire ice stones of varying sizes at various speeds at plywood decking panels covered with different samples of roofing materials. Their testing has shown that quarter-sized hailstones cause little to no damage, but the rate of damage increases dramatically with hail larger than 1". This is because the impact energy of a hailstone increases exponentially with the size of the ice ball, due to its heavier mass and higher velocity. For instance, a 2" hailstone striking head-on at a ninety-degree angle brings with it twenty-two times the impact energy than a 1" hailstone. Increase the hail diameter to 3", and that slightly larger-than-baseball-sized hailstone carries with it 120 times more impact energy than a quarter-sized piece of hail.

Hail forms as strong updrafts carry raindrops into the upper portions of the thunderstorm where they freeze and become ice pellets. These "hail embryos" may grow larger through a process of wet growth and dry growth that gives hail a layered look. In wet growth, the ice particle grows in an environment where there are many supercooled water drops. A supercooled droplet is one that has a temperature below freezing but remains in a liquid state. Through a process called riming, supercooled water drops collide with the small hailstone nucleus but do not immediately freeze. Instead, liquid water spreads across the tumbling hailstones and slowly freezes. Since the freezing process is slow, air bubbles are given time to escape, resulting in a clear layer of ice. If the updraft is strong enough, it can continue to keep the hailstones suspended in the cloud, allowing more time to encounter supercooled water droplets and therefore more time to grow.

Dry growth occurs when temperatures are well below freezing and supercooled water droplets immediately freeze upon contact with the small ice particles. In this case, air bubbles remain trapped within the ice and that icy layer takes on a cloudy or milky appearance. In the far upper portions of the thunderstorm where temperatures are extremely cold, hailstones may grow simply by fusing together with other pieces of ice, in a process called aggregation. Aggregation produces hailstones with a jagged appearance. When the hailstone becomes too heavy to be supported by the updraft or is blown out of the updraft, it falls to the earth. Generally speaking, the stronger the updraft, the larger the hailstones that can be produced by the thunderstorm. Recent studies using computer simulations of supercell thunderstorms also point to larger values of vertical wind shear, the change in direction and speed of the wind with height, as being favorable for producing bigger hail. The stronger vertical wind shear results in a broader updraft and a larger volume of the storm that is favorable for keeping hail suspended in the air, enabling the hail to grow larger.

MAYFEST HAILSTORM

On May 5, 1995, shortly after seven o'clock in the evening, a powerful super-cell struck Fort Worth. The area it covered included the Mayfest outdoor festival, along the banks of the Trinity River. More than ten thousand unsuspecting festivalgoers were caught without shelter as softball-sized hail pounded Trinity Park and the surrounding area. The freak storm created a nightmarish scene, as people either ran for their cars or huddled under anything they could find to shield them from the massive chunks of ice. Some who took shelter in their vehicles suffered lacerations from broken glass, as the hail smashed car windows and windshields. Fortunately, there were no fatalities at the festival, but at least sixty people were injured seriously enough to require hospitalization, and more than a hundred suffered minor injuries. Thirty of those injured were transported by city bus to Harris Methodist Hospital, where they were treated for lacerations and broken bones.

The Ridglea and Ridgmar areas were also pounded by the mega hail; the assault demolished cars and roofs. The glass portions of the roof at the Hulen Mall were shattered, showering mall patrons with shards of glass. Arnie Michelle was working at a store on the mall's second floor and says it was total chaos inside: "It sounded like a stampede on the roof. Glass was flying everywhere, water was pouring in, and there was so much screaming. We grabbed throw blankets off the display in our store and covered our heads to get downstairs." Luckily, she and her coworker escaped without injury.

The giant hail was the result of an HP (high-precipitation) supercell that developed suddenly near the apex of a fast-moving bow echo west of the Metroplex in Parker County. The radar images below show the supercell approaching the city of Fort Worth and then merging with the bowing squall line. The supercell moved into Tarrant County and lasted just long enough to ravage the city of Fort Worth with seventy-mile-an-hour winds and baseball- to softball-sized hail before it was overtaken by the bowing line

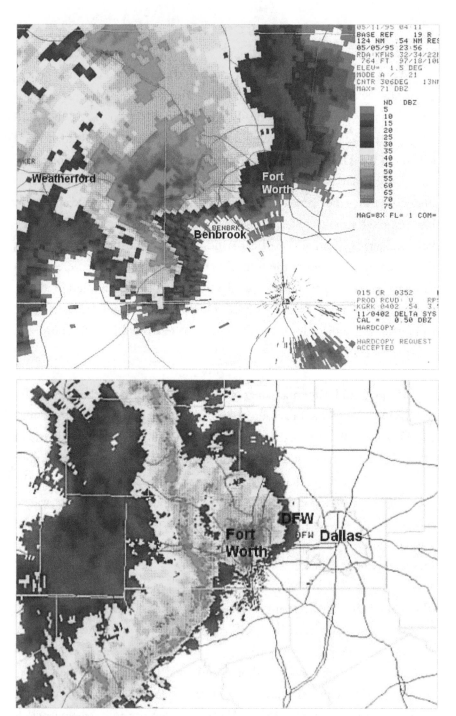

TOP: Radar Displays Large Hail Core (Pink) *(NWS Fort Worth)* BOTTOM: Radar Showing Squall Line Merging with Supercell *(NWS Fort Worth)*

of thunderstorms. Once the supercell was swallowed by the squall line, the complex of intense storms slowed down considerably and set the stage for a flooding catastrophe, which I will discuss in more detail in the next chapter.

The Mayfest storms resulted in $1.63 billion in insured losses ($2 billion total in storm damage), according to the Insurance Council of Texas. It is the costliest hailstorm in Texas history, just ahead of the April 2016 hailstorm that pummeled the San Antonio area with grapefruit-sized hail, causing $1.36 billion in insured losses. The National Weather Service (NWS) and local emergency managers were determined to learn from the mayhem at Mayfest, and, as a result, they changed the way they prepare for and collaborate during severe weather. In 1996, volunteers from the Radio Amateur Civil Emergency Service (RACES) began operating a mobile weather command center at Mayfest, equipped with radios and computers that connected them with the Fort Worth National Weather Service Office, emergency management, and city police in the park. They have trained weather spotters on-site and are in constant communication with NWS meteorologists, so they can clear everyone from the park in thirty minutes if severe weather threatens.

TRIO OF SUPERCELLS PUMMELS THE HEART OF THE METROPLEX

A single long-track supercell can be devastating, but on the evening of April 5, 2003, a trio of supercells tracked across the Metroplex, resulting in $1.14 billion in insured losses. The long-lived storms produced a nearly four-hundred-mile swath of damaging hail extending from near Snyder in West Texas to Texarkana. There were more than fifty reports of large hail from these three storms, with much of the damage occurring in a twenty-mile-wide area straddling the Denton/Tarrant and Collin/Dallas county lines.

This was my first bona fide severe weather event at FOX 4. I had arrived from Washington, DC only three months prior. Talk about baptism by fire! Meteorologists are essentially on call seven days a week during spring storm season. On this particular Saturday, my usual day off, I was shopping for a new TV and, on one of the big screens on display, I saw our weekend meteorologist, Ron Jackson, break into programming for a severe weather update. His update indicated the first storm was pummeling far western portions of the viewing area with softball-sized hail. Dragging me out of an electronics store is not easy, but one look at the rotating beast of a storm on radar convinced me I needed to hightail it into the station. It was going to be a busy evening.

That first supercell formed near the intersection of a stationary front and dry line near Big Spring, Texas, early in the afternoon. The stationary front ran from Big Spring across the entire state along the I-20 corridor and would serve as the "railroad tracks" for the giant hail-producing locomotives to roll into North Texas. As the afternoon progressed, the long-track supercell marched eastward. At 4:45 p.m., it produced a brief EF-0 tornado and grapefruit-sized hail in southern Throckmorton County. At 6:20 p.m., it produced another tornado with grapefruit-sized hail in northern Palo Pinto County. Continuing to roll along the stalled front, it moved into northwestern Tarrant County shortly after seven o'clock. Around the same time, a second supercell developed on the west flank of the first storm, followed by a third supercell that exploded roughly ninety minutes later. All three supercells took virtually the same path, traveling eastward through northern portions of Tarrant and Dallas counties and southern portions of Denton and Collin counties.

For the better part of four hours, the powerful storms battered highly populated areas with golf-ball- to baseball-sized hail, including one report of softball-sized hail in northern Rockwall County. The storms finally died out in Hopkins County after one final blast of golf-ball-sized hail at 10:50 p.m. At least three people were injured by the large hail that fell during this event, and thousands of homes, businesses, automobiles, and airplanes were damaged. The hail bombardment stripped countless trees and shrubs of all their foliage, leaving landscapes looking like it was the dead of winter.

Normally by the middle of June, the threat of big storms is over, but this June day would end with a barrage of giant hail rivaling anything we've seen in the heart of severe-weather season. Around five o'clock in the evening, a lone thunderstorm developed in the Richardson area. As the cell moved south, it tapped into a warm, humid air mass and intensified quickly. The maturing storm dropped hail that blew out skylights in Northpark Mall and then drifted south into East Dallas, where it pummeled homes and the historic Lakewood Theater with up-to-baseball-sized hail. The destructive hail continued as far south as Hutchins before the storm fizzled out. A second storm blew up over northwest Irving around seven o'clock, generating a swath of golf-ball- to baseball-sized hail as it churned from Irving and Grand Prairie southward through Cedar Hill and into Midlothian. Two additional supercells produced softball-sized hail and an EF-0 tornado near Leonard, along with eighty-five-mile-an-hour straight-line winds near Randolph that damaged homes, barns, and crops.

Meteorologists were late in identifying the potential for destructive hail and were not expecting to see baseball- to softball-sized hail. Out-of-season storms like these can be difficult to forecast. An important element in the storms' initiation was a well-defined thunderstorm complex that had moved east-southeast along the Red River and pushed a subtle outflow boundary into the Metroplex. We knew that the atmosphere was supportive of supercells, given the high instability, impressive wind shear with south-east winds at the surface turning clockwise to the northwest with height, and a cold pocket of air overhead. Despite the favorable shear, the lack of a distinct triggering mechanism to fire off the storms gave us reason to believe the severe weather threat was highly conditional. Obviously, we underestimated the boundary left behind by the thunderstorm complex, because it was effective in providing the spark to ignite the storms.

Dallas Lakewood Area *(Steve Reeves)*

The vicious storms left behind a trail of destruction. Tiled roofs of century-old homes on Swiss Avenue in East Dallas, thought to be nearly indestructible, were pulverized by tennis-ball- to baseball-sized hail. The salvo of hail was so loud it sounded to residents as if they were being bombed. The Lakewood Theater had its marquee and landmark tower completely stripped of its neon glass. Tens of thousands of cars throughout Dallas County had shattered windshields and windows as well as extreme body damage. Many vehicles were declared total losses, while scores of owners faced a wait of several months for repairs, due to the enormous backlog at body shops. The June 2012 hailstorm ranks as one of the costliest in Texas history, with total insured losses of $905 million.

APRIL 11, 2016, WYLIE HAILSTORM

If you search online for "hail size chart," you will discover that most charts top out at "softball-size" which is 4.5" in diameter. These indexes will suffice for 99.9% of hail, but on April 11, 2016, the city of Wylie in southeast Collin County was hit by an apocalyptic storm that hurled chunks of ice that were off the charts. One hailstone measured 5.25" in diameter and would officially become the largest recorded stone to ever fall in North Texas, although it was not a statewide record. Moore County (in the Panhandle) and Ward County (in Southwest Texas) have both been hit by hailstones that measured 6" in diameter. To put that into perspective, a size-three soccer ball used by children under age eight measures only slightly larger, at 7.25" in diameter.

A single, long-lived supercell that tracked east-southeast from near Bowie all the way to Van Zandt County was the culprit for the mega "hail-burst" that caused $300 million in damage. As the supercell moved into Collin County, it revealed its atmospheric might, towering to over 40,000'. The radar signature displayed an impressive bounded weak echo region (BWER), which is a column of very weak radar reflectivity bordered on the sides and top by significantly stronger radar returns. This indicates a very strong updraft; in this case, one that was likely traveling at speeds greater than one hundred miles an hour and was suspending much of the precipitation and hail in that portion of the storm. What goes up, must eventually come down, though, and when it did, Wylie felt the full fury of this powerful storm.

LEFT: Roof Punctured by Hail *(Michael Stidham)* RIGHT: Giant Hail from Wylie Storm *(Paul Stofer)*

The hail was so big and its impact so intense that it ripped gaping holes in rooftops, propelling chunks of ice right through many ceilings. Roughly 80% of the fifteen thousand homes in Wylie sustained damage that included pulverized roofs and siding, shattered windows, and severely shredded landscaping and tree foliage. In addition, all twenty campuses in Wylie ISD suffered damage, including its entire bus fleet.

JUNE 6, 2018, OVERNIGHT HAILSTORM

During the wee morning hours of June 6, 2018, an intense storm unleashed a torrent of hail that rudely awakened thousands of Metroplex residents. Unfortunately, this overnight hail event had not been forecast and caught many people by surprise. For much of the evening, our attention was focused on a cluster of storms that had developed near Amarillo and dove southeast. When that complex veered west of North Texas, the radar went silent and it appeared the rest of the night would be quiet.

The silence was broken by a rogue thunderstorm that developed in western Collin County just after midnight. It moved south for about an hour and then, shortly after one o'clock in the morning, the cell split, with the "right-splitting" portion heading toward Lavon Lake and the "left split" continuing south toward Lewisville. The storm moving toward Lewisville would cause most of the nocturnal ruckus. Around two o'clock, as it traveled south, it bombarded the Carrollton/Farmers Branch area with baseball-sized hail that smashed windows and pummeled roofs and cars. According to one Carrollton resident, "It sounded like someone was beating the roof with a baseball bat." The storm continued its rampage through Coppell, Las Colinas, and Arlington before fizzling out just after three o'clock. In its wake, it left twenty thousand damaged homes and twenty-five thousand damaged vehicles, the two-hour blitzkrieg having produced $1 billion of damage.

FIRE AND ICE

Lightning is a giant spark of electricity in the atmosphere between oppositely charged regions within clouds, the air, or the ground. The three main types of lightning are distinguished by where they occur: either inside a single cloud (intracloud), between two clouds, or between a cloud and the ground. While the process is not fully understood, research points to ice crystals, hail, and semi-frozen water drops called graupel or "soft hail" as critical players in "electrifying" the clouds by way of frictional charging. Lightning occurs due to a separation of electric charge and the generation of an electric field within a thunderstorm. As hail and graupel fall, they collide with much smaller supercooled water drops and ice crystals that are rising in updrafts. The resulting collisions strip negatively charged electrons off the ascending precipitation particles that collect on the descending particles. The falling hail and graupel acquire a negative charge and the rising supercooled water drops and ice crystals acquire a net positive charge. The result is a storm cloud with a negatively charged base and a positively charged top. A similar separation of charges occurs on a much smaller scale when we rub our shoes on a carpet. Electrons are stripped off the carpet and transferred to our body. The negative charge that develops is released as a spark as our finger reaches for a neutral or positively charged object like a doorknob.

Fort Worth Lightning *(Michael Beard)*

Air initially acts as an insulator or buffer, but as the difference in charges builds up, the air's insulating capacity breaks down. At this point, an invisible channel of electrons, called a "stepped leader," follows a zigzag path from the cloud base toward earth at a speed of roughly sixty miles per second. Next, a surge of positive charge called a "streamer" climbs upward through buildings, trees, and people into the air. Once contact is made between a streamer and a leader, usually a couple hundred feet above the ground, a complete conducting pathway is set, and lightning is set in motion. As many as a billion trillion electrons can transverse this path in less than a blink of an eye. The initial strike is often followed in rapid succession by several secondary strikes, but they are spaced so closely that it appears as a single, flickering strike. The massive, instantaneous flow of charge super-heats the air to a temperature of 50,000°F, causing the air column to expand violently. The expansion of air creates a shockwave that we hear as thunder. The flash of lightning temporarily balances the charged regions in the atmosphere until the opposite charges build up again.

TEXAS MAN SURVIVES A DIRECT LIGHTNING STRIKE

On October 3, 2019, Alex Coreas was at Meyer Park in Spring, just north of Houston, when a sudden thunderstorm hit the area. Alex was scurrying back to his truck with his three beloved German Shepherds in tow when he was struck by lightning. The terrifying ordeal was captured by a security camera at a nearby veterinary hospital. Video shows the bolt striking Alex, who immediately falls face first onto the pavement. The force of the lightning strike blew off his shoes and socks as it exited through his feet and left a gaping hole in the concrete parking lot. Fortunately, several Good Samaritans, including a vet technician, rushed to his aid and immediately began performing CPR. After multiple rounds of chest compressions, they were able to get a pulse, and Alex was breathing again when the ambulance arrived. Alex was airlifted to a local hospital where he spent four days receiving treatment for fractured ribs, a fractured temporal bone (behind his right ear), a badly swollen eye, and multiple muscle ruptures. He also suffered burns from his chest, where his shirt had caught on fire, down to his left foot. While Alex did lose hearing in his left ear, he considers himself lucky to have survived.

THE FLASHOVER EFFECT AND WHAT TO DO WHEN THUNDER ROARS

Of every ten people who get hit by lightning, nine will survive. This is because the vast majority of lightning strike victims do not get hit directly but rather are struck by "side flashes." Lightning strikes a nearby building or tree, for example, and there will be a side flash, carrying a small fraction of the original bolt's current, that jumps through the air to the person. Alex experienced a direct strike, and he may have survived due to the fact that the human body tends to resist the flow of electric current. Only a tiny amount of electricity ricochets through the body. Most of it travels around the outside surface of the body in an arc from head to toe and bypasses the vital organs, a phenomenon called the "flashover effect." Many lightning strike survivors still suffer a variety of short- and long-term effects, however, including seizures, deafness, headaches, memory loss, personality changes, and chronic pain.

The odds of becoming a lightning strike victim in the United States in a given year are roughly one in seven hundred thousand. Assuming an average lifespan of eighty years, a person's odds over their lifetime become one in nine thousand. What can you do to avoid becoming a statistic? Take shelter as quickly as possible, preferably at the first sight of storm clouds building. If you hear thunder, no matter how faint, you are close enough to be struck by lightning. Go indoors, or to your vehicle if there is no safe shelter nearby. Finally, wait at least thirty minutes after the last rumble of thunder to return outside: 50% of lightning deaths occur after the storm has passed.

STRAIGHT-LINE WINDS CAN PACK A POWERFUL PUNCH

Tiphannie Willie-Clements and her ten-year-old son were fast asleep early on the morning of March 29, 2017, when a vicious storm hit. "The house started shaking and rumbling and within five seconds it felt like it exploded," Tiphannie recounted in an interview with a FOX 4 reporter just a few hours later. Sky 4 aerial images of the extreme damage in her Dalton Ranch neighborhood in Rockwall showed entire second floors ripped off of two homes and much of the front facade of a third home peeled away, with portions of the house's roof missing. The powerful blast of winds was not produced

Rockwall Damage *(KDFW-FOX 4 News)*

by a tornado, but by a bowing line of thunderstorms that generated ninety-five-mile-an-hour straight-line winds, the equivalent of an EF-1 tornado. National Weather Service meteorologist Mark Fox, who surveyed the damage, describes what he believed happened: "The garage doors were the first thing to fail on these homes. Once the wind got into the house, it tried to get out of the house as quickly as possible and that's why the roofs came off." Two dozen other homes in the Rockwall subdivision suffered damage.

Severely Damaged Home *(DFW Red Cross/Mark Bishop)*

The term "straight-line winds" applies to any thunderstorm winds that are not associated with rotation. These winds occur much more frequently than tornadoes and can unleash as much power as an EF-0 or EF-1 tornado. Downbursts, which are powerful winds that plunge from a thunderstorm and spread out quickly once they "splash" down, are often cited as the source of these winds. The most intense downburst on record happened in August 1983, at Andrews Air Force Base, just outside of Washington, DC. A peak wind gust of 149 miles an hour was measured at 2:11 p.m., just seven minutes after *Air Force One*, with President Reagan on board, had landed on the same runway where the extreme gust was recorded.

SUNDAY STORMS TURN DEADLY IN DALLAS

On June 9, 2019, a compact but potent east–west oriented line of severe thunderstorms rumbled southward out of Oklahoma in the morning and intensified as it rolled through North Texas in the early afternoon hours. Complexes of thunderstorms like this one remain well organized by sustaining a "cold pool" of air that acts like a cold front to continually generate new thunderstorm development. The rain-cooled air that forms the cold pool descends from the mid-levels of the atmosphere to the ground on both sides of the advancing line of thunderstorms. The phenomenon is called a downburst, and the June 9 downburst produced a broad belt of intense, straight-line winds that moved southward in tandem with the thunderstorms as they forged ahead. Winds were clocked at sixty-four miles an hour in Corinth in Denton County at 1:08 p.m. and sixty-seven miles an hour at Rhome in Wise County at 1:32 p.m. As the powerful storms moved into Dallas County, they intensified. Dallas Love Field recorded a seventy-one-mile-an-hour wind gust at 1:45 p.m., and there were several other estimates of seventy- to eighty-mile-an-hour gusts reported by storm spotters as the complex roared south from the Metroplex through Ellis and Navarro counties. Hurricane-force gusts downed trees and power lines, which blocked the streets and triggered massive power outages. The violent windstorm turned deadly when a crane collapsed on the Elan City Lights Apartment Building in Old East Dallas. The crane sliced through the five-story apartment complex, killing a resident and injuring five others. The entire building was condemned, and its five hundred residents were forced to find a new home.

LEFT: Dallas Crane Collapse *(KDFW–FOX 4 News)* RIGHT: Carrollton Fire Search and Rescue Inside Crushed Building

The self-perpetuating nature of these thunderstorm complexes is linked to the cold, dense air they generate. If the cold pool races ahead of the squall line, it cuts off the warm, moist supply of air feeding the storms, and the squall line weakens and eventually falls apart. In this case, there was a balance between the advancing cold pool from the north and the warm, moist flow directed toward the squall line from the south. The convergence of the two air masses thrusted the air upward and continually generated new cells along the leading edge of the squall line. Both of these events are prime examples of why severe thunderstorm warnings need to be taken seriously.

A TEXAS-SIZED WINDSTORM

On the afternoon of May 4, 1989, an isolated supercell developed over southeastern Colorado and moved southeast into the Amarillo area after five o'clock in the evening. As it continued its southeastward jaunt, it began to evolve into a larger bow echo. When the line reached the Dallas–Fort Worth area around midnight, it had expanded into a formidable bowing complex of storms well over one hundred miles in length. The giant bow echo generated powerful winds that gusted to ninety-five miles an hour at a power plant near Graham and one hundred miles an hour in Fort Worth. The intense winds continued to rage as the line of storms plowed through

Deep East Texas overnight and into southeast Louisiana around seven the next morning before finally dissipating.

The widespread, long-lived windstorm, called a "derecho" (deh-REY-cho), damaged thousands of businesses and homes and destroyed more than a hundred structures in North Texas, many of which were mobile homes. Some of the worst damage occurred in Young, Palo Pinto, Hood, Tarrant, and Johnson counties. Two people were killed, and several dozen others injured. Derechos can be associated with a single large bow echo or a squall line with multiple bowing segments that repeatedly produce clusters of downburst winds as they move along. These serial wind generators often travel at speeds of fifty miles an hour or greater and have been known to produce wind gusts as high as 130 miles an hour. Embedded circulations within these well-developed storm complexes may also produce tornadoes.

DOWNBURSTS: MICROBURST TO MACROBURST

In the lead-up to a downburst, a large core of rain, hail, and graupel that had been buoyed by the updraft begins falling toward the ground. As it falls, it drags a great deal of air along with it and gains momentum as it continues to plunge. If the air beneath the storm is relatively drier, some of the rain evaporates and cools the air further, making the bundle of air and precipitation heavier and increasing its speed. When the downdraft hits the ground, it acts much like water pouring from a faucet and hitting the sink bottom. The water stream is the downdraft, and the outward spray, which spreads out in all directions, is the downburst. A macroburst is an outward burst of strong winds at or near the surface, with horizontal dimensions larger than four kilometers (two and a half miles), sometimes producing damage similar to that of a tornado.

A "microburst" is the most commonly encountered type of downburst. It is small in scale, less than two and a half miles in width, and generally lasts five minutes or less. Microbursts fall into two categories: dry and wet. Dry microbursts, common in the High Plains and the Western United States, occur with little or no precipitation reaching the ground. I have witnessed these most often during the summer on trips through West Texas. One of the visual cues I look for are dark streaks of virga (precipitation evaporating before reaching the ground) below high-based, well-developed thunder-

Arizona Wet
Microburst
(Bryan Snider)

storms. This is a sign that the downdraft may acquire substantial momentum due to evaporational cooling, which will lead to very gusty winds upon impact at the ground. Wet microbursts, more common in the Midwest, the East, and the Southern United States, are accompanied by heavy precipitation at the surface. To understand why wet microbursts are popularly referred to as "rain bombs," watch the video.

DELTA FLIGHT 191 TRAGEDY AT DFW AIRPORT

On August 2, 1985, Delta Airlines Flight 191 had a scheduled stop at DFW Airport at 6:05 p.m., en route from Fort Lauderdale to Los Angeles. The

Ian and Richard Laver

302-seat jet was carrying 152 passengers and eleven crew members. Among those on board were twelve-year-old Richard Laver and his father, Ian, a cousin of tennis great Rod Laver. They were on their way to a tennis tournament in California. The night before the Lavers boarded their flight, Richard told his mother of a bad dream that was haunting him. "I told her, I really have a bad feeling we're not going to make it tomorrow on this flight." My mom said, "Wow, where is this coming from?

You've flown all over the world." Then she said, "It's a one in a million chance. That'll never happen."

Richard and his father nearly missed their flight the next day. After take-off, Richard's anxiety temporarily eased as the first leg of the trip began with blue skies overhead. Most of the flight was smooth. However, upon approach to DFW, Richard saw ominous clouds out his window and was overcome with fear. He went to the bathroom, where he splashed water on his face to try to regain his composure, but he was convinced they were going to crash. "As I walked back to my seat, I remember thinking, 'Don't put your seatbelt on.' So I took the blanket and put it over my lap so the stewardess would not see I wasn't wearing it. Ultimately, that's what saved my life...everyone [else] in my row died."

THE FINAL TWO MINUTES BEFORE IMPACT

Just three miles ahead of Flight 191 was a Learjet on the same approach to runway 17L. The Learjet flew through a thunderstorm just north of the runway, hitting heavy rain, but reported only "light to moderate turbulence" and landed safely. The flight crew of Delta 191 encountered the same storm on their approach to runway 17L. Two minutes before the landing gear would make first contact with the ground, the captain radioed the tower and reported, "Tower, Delta 191, heavy out here in the rain, feels good." The tower controller advised the crew that the wind was blowing at six miles an hour with gusts up to seventeen miles an hour, and the flight crew lowered the landing gear and extended their flaps for landing. Ninety seconds before touching down, the first officer commented, "Lightning coming out of that one. Right ahead of us." For the next sixty seconds, Flight 191 continued its approach without any issues. Then, just over thirty seconds before first ground contact, their speed increased from 171 to 199 miles an hour as they entered the leading edge of the microburst. The captain, now recognizing that the sudden increase was a result of wind shear, warned his first officer, saying, "You're going to lose it all of a sudden, there it is." Moments later, their air speed dropped abruptly to 153 miles an hour. At the same time, the cockpit voice recorder captured the sound of heavy rain as they entered the core of the intense microburst. Weather observations revealed that the microburst produced wind gusts over eighty miles an hour. Over the next thirty seconds, the crew struggled to regain control as the jet pitched sharply from right to left, but violent swirling winds, likely vortex rings, and strong tailwinds made it impossible to regain altitude. Just seven seconds before ground contact, the captain declared "TOGA" (take off/go around), aviation shorthand meaning to apply maximum thrust and abort the landing, but they could not stop the plane's descent.

At 6:05 p.m., the plane landed hard in a field just over a mile north of the runway. Still structurally intact and rolling at a speed of over two hundred miles an hour, the wide-body jet skidded across Highway 114, shearing off tall highway lights and striking a vehicle. The impact killed the driver instantly. With flames emerging from the left wing, the aircraft careened across an open field on the edge of the airport property and then struck two water tanks. The impact with the second tank caused the aircraft to split open and become engulfed in a massive fireball. Richard's memory of the final moments of the flight has only recently started to come back to him. "The plane went through crazy amounts of turbulence. It felt like a crazy drop. Like an elevator dropping. It felt like my dad was covering me, trying to protect me. Then I saw a flame come at me and I hear this god-awful noise. The next thing I know, I am in a field and I am freezing cold." Richard vividly recalls feeling the pain of the hail pelting his face, which had been burned by the flames. Thousands of gallons of water were gushing out of the storage tank that had been torn open by the aircraft. "I'm gargling water and spitting it out. I thought I may be in danger of drowning." In shock, Richard tried to cry out for help but could not utter a word. The storm was still kicking up violent wind gusts and there were small explosions going off in the distance. "Then I felt this anger inside of me that said, 'I will not die here today.'" Determined to live, his voice came alive and he cried out to anyone who could hear him. Soon after, Richard was pulled from the flooded field by a passing motorist who had stopped to look for survivors. Richard remembers his rescuer telling him, "You're going to be okay."

Delta 191 Wreckage *(KevinBrownPhoto.com)*

The crash of Delta 191, still the worst in Texas aviation history, killed 126 of the 152 passengers on board, eight of the eleven crew members, and one motorist on the ground. The fuselage from the cockpit to row 34 disintegrated, while the tail section skidded backwards away from the intense flames. Richard was sitting in the first row of the charred tail section and was ejected when the plane broke apart. Most of the twenty-six survivors were in the tail section that broke away from the rest of the aircraft.

HOW A MICROBURST DOWNED A JETLINER

In its report, the National Transportation Safety Board concluded that the crash was a result of the crew encountering microburst-induced severe wind shear at low altitude. To better understand what happened, scan the QR code below to watch a simulation of Flight 191 as it flew into the microburst. As the plane first enters the microburst, it faces strong headwinds, pushing more air over the wings, which produces a marked increase in lift and a gain in altitude. To adjust, the pilot pushes the nose down, but then the aircraft flies through the downdraft and suddenly the headwind shifts to a tailwind. With much less air now moving over the wings, lift is greatly reduced, and the aircraft rapidly loses altitude. This is precisely what had happened to Delta 191 when the captain is heard on the cockpit voice recorder saying, "Push it up, push it way up." Despite the crew's best efforts, the swirling outflow winds made regaining control and altitude impossible and forced Delta 191 to the ground.

Delta 191
Microburst
Animation *(FAA)*

THE LEGACY OF DELTA 191— MAKING THE SKIES SAFE

Dr. T. Theodore Fujita, a pioneering severe-weather researcher who developed the original Fujita Scale for tornado intensity, coined the term "microburst" in 1975. While investigating the fatal crash of Eastern Airlines Flight 66 at JFK Airport in New York, he described a microburst as a wind system strong enough to bring down a jet aircraft but small enough to go undetected by our network of surface weather observations. This discovery led him, in 1978, to conduct the first major field experiment that used Doppler radar to successfully detect microbursts: the Northern Illinois Meteorological Research on Downbursts (NIMROD). Soon after, the Federal Aviation Administration (FAA) began installing low-level wind shear alert systems (LLWAS) in the late 1970s and early 1980s. However, these first LLWAS had high false-alarm rates and their sensors were too far apart to detect intense microbursts like the one that brought down Delta Flight 191.

The Delta 191 crash was the wake-up call that prompted the federal government to accelerate its efforts to improve safety for the flying public. In 1988, the FAA ordered all commercial aircraft have onboard wind shear detection systems installed by 1993. Many airports across the United States, including DFW, also installed ground-based wind-shear detection equipment.

Today, DFW has a network of eighteen LLWAS sensors along with a terminal Doppler weather radar that probes the skies. Pilots undergo intensive training programs that include "flying the DFW microburst" in a simulator. With modern wind-shear detection systems in place and flight crews receiving improved training, downburst-related accidents have radically declined, and air travel is much safer.

SAVING KATE AND THE "LITTLE BOY IN THE FIELD"

Richard Laver survived the crash of Delta Flight 191, but his life truly wouldn't be saved until thirty years after being plucked out of the muddy field at the end of runway 17L. Long after his physical injuries had healed, Richard was traumatized by visions of planes crashing into the building where he worked. When he heard or saw a truck driving down the freeway, it reminded him of a runway. The mere sight of an airport would trigger post-traumatic stress disorder symptoms. "The world became a very unsafe place for me," says Richard. For many years, his life had no direction and no purpose. He was even homeless briefly. Then he met his wife, Michelle, in 2003, and she helped to restore his hope and gave his life new meaning. In 2006, they welcomed their miracle baby, Kate. Kate was diagnosed at birth with cerebral palsy. Kate's condition made eating very difficult. She was prescribed various formulas, but they all made her sick and were not providing her adequate nourishment. At five years old, she weighed a mere sixteen pounds. Desperate to get Kate the proper foods she needed to nourish her frail body, Richard worked with a vegan chef, nutritionists, and dieticians to create a healthy meal-replacement shake. The team researched and gathered together thirty superfoods loaded with vitamins, powerful antioxidants, and plant-based proteins, mixed them in a blender, and began feeding the shakes to Kate. It worked wonders, Richard says. "Within four to six weeks, Kate was off every medication and she was gaining tons of weight. She was finally well again."

The Lavers soon ran into a problem. California law did not permit students to bring non-packaged foods that were not approved by the FDA to school.

Richard knew what he had to do but taking his kitchen-recipe shake to market and preparing it to be commercially packaged would require him to risk everything. The odds didn't matter to Richard. After all, this was a man who had survived a plane crash as a boy. "It took two years and 86 batches to get it shelf stable (safely stored at room temperature in a sealed container for six months). I spent all my money, down to my last $5,000." Faith and perseverance eventually paid off. In 2013, the first commercially sealed bottle of Kate's shake hit the shelves, and the team's company, Kate Farms, was born. To date, Kate Farms products have nourished tens of thousands of people around the world, many of them with medical conditions like Kate. Richard believes that he has finally overcome his demons and has found an inner peace in his life. "What I learned in therapy was that I spent the rest of my life after the crash protecting that little boy in the field, and that by saving Kate, I was saving that little boy in the field."

CHAPTER 8

FLASH FLOODING, HURRICANES, AND TROPICAL DELUGES

B efore flood-control measures were implemented to help tame the Trinity River, the cities of Dallas and Fort Worth were frequently swamped by flood disasters of biblical proportions. These were not floods that brought two feet of water and forced a few lanes of highway to close during rush hour. These were floods that swallowed the first floor of buildings throughout vast portions of downtown and rushed, unremitting, splashing the rooftops of entire neighborhoods. One such flood occurred in May 1908. Over a three-day period, North Texas was inundated by 10" to 15" of rain, which resulted in a flood gauge reading in Dallas of 52.6'. The swollen Trinity River stretched two miles wide between West Dallas and downtown Dallas and caused millions of dollars of damage. The entire city of Dallas went dark for three days; it had no telephone, telegraph, or rail service; and Oak Cliff could only be reached by boat. The flooding killed eleven people, and four thousand others fled their homes to seek higher ground. Fourteen years later, in 1922, another flood disaster struck when the Trinity broke through levees and swamped the city of Fort Worth, leaving thirteen dead and a thousand homeless, and causing at least $1 million in damage. History repeated itself yet again when Fort Worth suffered what may have been its worst flood ever in 1949.

On the night of May 16, 1949, up to a foot of rain fell on the Upper Trinity River Basin. The Clear Fork of the Trinity River burst through its levees, releasing torrents of floodwater into the city. The area of the present-day West 7th shopping district was besieged with 10' to 12' of muddy water, while farther southwest, the Fort Worth Zoo and Colonial Country Club were submerged under 6'.

The West 7th and Lancaster Street bridges (in the upper left of the aerial photo below) were closed, as houses floating down the river smashed into the bridge pilings. A line of utility poles poking out of the water in the center of the photo marks White Settlement Road, while flat roofs are the only visible portion of the commercial buildings in the foreground. Will Rogers Coliseum, the stock show cattle barns, Carswell Air Force Base, schools, and churches all became shelters for the estimated thirteen thousand people who fled their homes. Ten people were killed, and property damage was estimated at $11 million, the equivalent of over $100 million today.

LEFT: 1949 Fort Worth Flood *(The Portal to Texas History/Lockheed Martin)* RIGHT: Montgomery Ward Building *(City of Fort Worth)*

Congress reacted to the 1949 flood by commissioning a new Army Corps of Engineers district in Fort Worth in 1950 to carry out levee flood-control projects in Dallas and Fort Worth, and reservoir projects such as Grapevine Lake (1952) and Lake Lewisville (1955). In addition to creating lakes and reinforcing levees, the Army Corps of Engineers completed a floodway reconstruction project to deepen and widen the river channel in 1958.

From the 1960s to the 1980s, an explosion of growth and development in the Dallas-Fort Worth Metroplex incrementally increased the runoff produced by the Trinity River watershed. Several major flood events from 1989 to 1991 brought runoff that exceeded the capacity of flood-control reservoirs, and the resultant flooding caused hundreds of millions of dollars in damage from Eagle Mountain Lake to South Dallas. Design engineers for Lake Benbrook in Fort Worth, completed in 1952, had estimated that waters in the flood-control reservoir would reach the 710' elevation of the spillway only once every forty years, and that elevations of 715' or greater would be reached only once every hundred years. Yet lake levels reached 717' in May 1989, 718' in May 1990, and 713' in December 1991, contributing to widespread flooding downstream in Dallas. Since then, miles of flood-control wetlands have been constructed, new levees have been added and existing ones improved, and state-of-the-art pumping stations have been built, which, when combined, can siphon millions of gallons of stormwater per minute, directing it away from residential and business districts during heavy rain events.

In 2018, Governor Greg Abbott announced that $5 billion in federal funds would be allocated for flood-control projects in Texas; $370 million of that was earmarked for critical flood-control projects along the Trinity River. These projects include two new pump stations, construction of a new levee south of downtown Dallas, and desperately needed repairs and upgrades to the sixty-three-year-old Lewisville Lake Dam. A breach of the dam would be catastrophic for the three hundred thousand residents living downstream, as well as the $14 billion worth of property in harm's way. While the twenty-first-century infrastructure will certainly help contain the Trinity River and protect the cities of Dallas and Fort Worth from the monumental surges witnessed prior to 1950, another flooding threat looms large, and we have seen that it can strike fast and furious, with deadly consequences.

FLASH-FLOODING NIGHTMARE

When Mark and Joy England put their three toddlers to bed on the night of September 21, 2018, it was sprinkling outside their Red Oak residence, about twenty miles south of Dallas. The Englands, including a teenage son and daughter, lived in a one-story, ranch-style home on an acre that bordered Red Oak Creek. Through much of the year, the creek runs only a few inches deep, meandering gently along the edge of their property. Late that evening, a complex of thunderstorms slowly pinwheeled into North Texas from the west, and the evening sprinkles turned to tropical downpours that continued unabated for hours.

At about two o'clock in the morning, Joy and Mark awoke to a trickling sound in their bedroom. Mark got out of bed to investigate the cause and immediately felt that the floor was wet. Moments later, curiosity turned to deep concern when Mark turned on the outside light to see water lapping against the house, nearly halfway up the French door. The couple knew they had little time to act. As Mark looked for a safe way out, Joy went room to room to wake the kids. "By the time I got back to our bedroom, the French door started busting open by the power of the water. The water was now knee deep in our house. At that point I thought, 'We need to get out now!' So I went back to the toddlers' room and their beds were now floating."

Mark and their teenage son, Caleb, tried to get out through the front door, but the enormous pressure of the water had jammed it shut. Their only option was to go out through the front window. Wearing only their pajamas, the family climbed through the window and dropped down into chest-deep water.

"We thought we would just swim up the driveway to get to higher ground. The water was already halfway up the van. Then we watched our daughter's car float off down the creek." The water was whirlpooling and creating a strong, swift-moving current that was blocking their escape. "Mark grabbed our teenage daughter, who had the two-year-old strapped on her front and the four-year-old on her back, and the little dog. I don't know how he did it, but he literally pulled them through the chest-high current, out to a tree near the main road, and they held on." Meanwhile, Joy, her three-year-old, and her teenage son floated through the front yard to their zipline platform. The water was rising quickly and was just below the fifteen-foot platform, which was attached to a tree. As Mark swam back through the raging current to assist them, Joy called for help. "The rain [was] pouring down. 911 couldn't hear us. [I was] yelling our address as loud as I [could]." The emergency dispatcher managed to figure out their location and send help, but then told them that first responders could not reach them. The two bridges that served as the only access points to their neighborhood were blocked by huge trees.

Joy says that was the moment when they realized they were going to die unless they took matters into their own hands. Mark hatched an idea. If they swam hard, they could ride the floodwaters to an abandoned house next door. Once they reached it, they would break in through a window and go to the second floor to get above the water. "So Caleb and I took off swimming and Mark followed us with [our daughter] on his head. Right...as the current is

TOP: Red Oak Creek Flooding 9/22/18 *(Maverick Drone & Photography Services)*
BOTTOM: England Family *(Joy England)*

about to smack us into the house next door, we started touching grass with our feet. We couldn't believe it. We were so excited. 'We're touching! We're touching!'" From there, they found enough footing to move uphill to dry ground and reunite with their daughter and the other two children. By six in the morning, just three hours after the creek rose 20', the water had receded.

FLASH FLOODING AND ITS UNIQUE FORECASTING CHALLENGES

According to the American Meteorology Society (AMS), flash flooding is a "flood that rises and falls quite rapidly with little or no advance warning, usually as the result of intense rainfall over a relatively small area." In the case of the England family, the water level on Red Oak Creek rose 20' in no time at all. Flash flooding presents a unique warning challenge because forecasters must take into account more than just meteorological factors. Unlike other severe-weather events, such as tornadoes and hail, flash flooding can be caused and aggravated by a number of contributing elements. The Englands lived in a more rural portion of Ellis County, but highly populated areas may be at greater risk for flash floods owing to a dense network of highways and roads and a multitude of other impervious ground cover, including thousands of acres of parking lots and paved surfaces. Natural surfaces are very good at absorbing rainfall, but when water hits pavement, it creates runoff immediately.

Tom Bradshaw, Meteorologist in Charge at the National Weather Service Forecast Office in Fort Worth, says explosive development in North Texas has made us highly vulnerable to flooding. "We have so many urbanized areas here with so much impermeable concrete, so the ability for water to run off in rapid fashion and reach flash flood proportions is so much greater than it was twenty-five to fifty years ago." Consider a one-mile stretch of highway with four westbound and four eastbound lanes plus a service road running adjacent to the highway in each direction. Just 2" of rainfall on that one mile stretch of paved surfaces would produce one million gallons of water, an enormous amount of runoff to capture and drain. Fort Worth alone has over 7,500 miles of street surface and only $10 million annually to spend on stormwater-runoff improvement projects. That may sound like a lot of cash, but it's a drop in the bucket when you consider the official list of stormwater-system improvements that would require the city to spend close to $1 billion. Bradshaw says that, for many cities, it's simply not in the budget: "It's very expensive for communities to make the investment in infrastructure needed to effectively manage stormwater." Exacerbating the problem are construction zones and low-lying areas under overpasses and bridges where the surge of runoff can easily overwhelm the drainage system and result in rapid water rises that can quickly become death traps for motorists.

TEXAS LEADS THE NATION IN FLOODING DEATHS

While tornadoes and hurricanes often dominate the weather headlines, the leading weather-related killer almost every year across the country is flooding. Over the past thirty years, flooding has claimed more Americans' lives than tornadoes, hurricanes, and lightning. Sadly, Texas is a major contributor to those statistics. From 2010 to 2018, there were 212 deaths due to flooding in the Lone Star State, more than triple the number of fatalities in Arkansas, which has the second-highest number of fatalities, at sixty-three. Though Texas is a large state, geographical area alone does not justify these inflated numbers. The Houston area accounts for a disproportionate number of fatalities. The "Bayou City" is vulnerable to tropical downpours from the remnants of hurricanes and tropical storms and to slow-moving clusters of heavy thunderstorms. Harris County, which encompasses Houston, averages five days of flooding each year. Houston's streets serve as part of its flood-control system, emptying stormwater into its ten bayous, which flush the runoff into Galveston Bay on the way to the Gulf of Mexico. Since Houston only sits at 43' above sea level, the flat system of meandering bayous empties slowly.

Claiming five of the top six highest rainfall amounts from tropical cyclones in the contiguous United States, Southeast Texas is without question one of the most flood-prone locations in the country. The latest to leave its mark was a storm named Imelda that caught many in its path by surprise in September 2019. Imelda dumped 43.39" of rain at Fork Taylors Bayou, located fifteen miles southwest of Beaumont. It demonstrated that even a weak storm can have a catastrophic impact; Imelda became a tropical depression just three hours prior to landfall and barely reached tropical storm strength an hour before coming ashore near Freeport.

In addition to Imelda, Amelia (1978), Claudette (1979), and Allison (2001) all produced rainfall amounts in excess of 40" in Southeast Texas. Topping all of those storms, Hurricane Harvey became the wettest tropical cyclone on record in the United States when it unleashed over 5' of rain (60.58") in Nederland, Texas, in 2017. The floodwaters damaged or destroyed at least 200,000 homes and businesses, displaced more than thirty thousand people, and prompted more than seventeen thousand rescues. To better understand why tropical storms and hurricanes produce five-hundred-year floods on a regular basis in Texas, let's first take a look at how they form and how they generate such destructive power.

Hurricane Harvey Rescue *(Army 1st Lt. Zachary West)*

HURRICANE FORMATION

Hurricanes have humble beginnings, born as clusters of thunderstorms over warm ocean waters, and nurtured by a vast supply of water vapor. Most of these tropical disturbances encounter hostile winds or dry air that disrupt their organization and quickly lead to their demise, but roughly ten tropical disturbances in the Atlantic Basin grow into tropical storms each year, and six of those mature into full-fledged hurricanes. That maturation process requires just the right mix of atmospheric ingredients:

- Sea surface temperatures of at least 80°F
- An atmosphere that cools quickly enough with height for thunderstorm development
- Moist air through a deep layer of the atmosphere
- Light winds from the surface to the top of the troposphere (roughly eight miles up)

With intense heating from the tropical sun, ocean moisture is evaporated into the drier air above it. The energy required to convert the liquid to water vapor lies dormant or "latent," waiting to be released when the vapor condenses into liquid again. This occurs with rising air in clouds and thunderstorms. As more thunderstorms erupt, tremendous amounts of latent heat are released as the ocean moisture condenses and the surface pressure drops. In response to the falling pressure, the air begins to converge more quickly toward the center of the low. As the air moves closer to the center, its speed increases, just as ice skaters spin faster as they bring their arms close to their bodies (conservation of angular momentum). These swifter winds kick up turbulent eddies (circular swirls) that transfer the warm, moist air upward where the water vapor condenses to fuel more thunderstorms. The cycle continues to repeat itself, allowing more thunderstorms to blossom and the storm to become better organized.

The process of hurricane formation mirrors the mechanics of a heat engine. In a heat engine, heat is taken in at a high temperature, converted to mechanical energy, then released at a lower temperature. In a hurricane, heat is drawn from the warm ocean surface, converted into wind energy, and then released in a cool exhaust in the uppermost portions of the storm. How strong is the engine that powers a hurricane? Incredibly, the energy released by condensation in an average hurricane in a single day is at least two hundred times the entire world's electrical energy production capacity.

STAGES OF DEVELOPMENT

A hurricane goes through stages during its life cycle, beginning as a tropical disturbance. When the cluster of thunderstorms develops a closed wind around a well-defined center, the tropical disturbance becomes a tropical depression. If the tropical depression becomes better organized and winds increase to thirty-nine miles an hour, it is upgraded to a tropical storm and, at that point, becomes significant enough to acquire a name. Tropical storms resemble hurricanes due to the intensified circulation. As the surface pressure continues to drop, a tropical storm becomes a hurricane when wind speeds ramp up to seventy-four miles an hour.

At this stage, the storm has a pronounced rotation around its central core and may span hundreds of miles across. The strong winds blowing over long stretches of the ocean cause a huge mound of water to pile up near the eye of the hurricane, which can cause a mammoth storm surge when this accumulated water reaches the shore.

Hurricanes are classified on a scale, called the Saffir-Simpson scale, which ranges from 1 to 5. The higher the number, the greater the wind speed and the more potential for damage. A storm is considered a "major" hurricane when it reaches the category 3 stage. A major hurricane generates destructive winds over 111 miles an hour and a storm surge of up to 12'. Category 5 hurricanes pack winds in excess of 155 miles an hour and a devastating storm surge that can easily top 20', depending on the slope and shape of the shoreline. Hurricane Dorian struck the Abaco Islands on September 1,

2019, with maximum sustained winds of 185 miles an hour, tying the Labor Day hurricane of 1935 for the highest winds of an Atlantic hurricane ever recorded at landfall. Dorian went on to strike Grand Bahama at similar intensity, stalling just north of the territory with unrelenting winds for the better part of forty-eight hours. Only five category 5 hurricanes have made landfall in the United States.

ANATOMY OF A HURRICANE

The tropical showers and thunderstorms associated with hurricanes are organized into bands (spiral rain bands) that swirl in toward the storm's center and wrap around the eye. Surface winds increase in speed as they spiral toward the center. In the diagram, the red and orange arrows show winds blowing counterclockwise into the hurricane at the surface, and blue arrows show winds spiraling outward, clockwise, at the top of the storm. The clockwise flow in the uppermost portion of the hurricane serves as the cool exhaust for this highly efficient atmospheric engine.

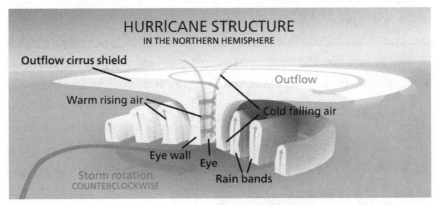

Cross Section of a Hurricane *(Wikimedia Commons/Kelvinsong)*

Circling the eye is the eyewall, a ring of intense thunderstorms where the hurricane's full fury is felt. Fierce winds whip up blinding sheets of rain, falling at rates of up to 15" per hour. Within the eye of the hurricane, winds suddenly die down, the rain ceases, and sinking air produces relatively clear skies. It's not unusual to see the eye corral flocks of birds during the day and reveal moonlit skies at night. Hurricane hunters, who fly reconnaissance missions to gather critical data inside the storm, often witness one of the most awe-inspiring sights in nature. As they punch through the eyewall and enter the eye, they find themselves surrounded on all sides by towering walls of cumulonimbus clouds that soar to heights in excess of 50,000'. Because the clouds of the eyewall tend to slope outward and up from the center, an individual in the eye of a hurricane has the impression of being in the bowl of a large stadium; this appearance has been dubbed the "stadium effect."

SUPERCHARGED STORMS ARE SWAMPING SOUTHEAST TEXAS

Authors of several studies have pored over data to gain insight into these "supercharged" storms like Harvey and Imelda. They cite two primary reasons for the epic flooding in recent decades: slower storm movement and higher ocean heat content. One study was led by NOAA hurricane researcher James Kossin. Kossin examined the tracks of all tropical cyclones (hurricanes and typhoons) across the globe from 1949 through 2016. His number-crunching revealed that the speed of tropical cyclone movement has slowed by 10% over the oceans and by 20% over land compared to historical averages. Kossin claims the more leisurely pace is tied to climate change. As

Imelda Flooding *(Splendora, TX/Jamesa Larimer)*

the earth's atmosphere warms, the atmospheric circulation changes. As a result, the steering winds that move these storms along have become weaker. The lumbering pace of tropical systems means they spend more time over land, resulting in unprecedented rainfall totals. Both Harvey and Imelda moved at a snail's pace. Harvey crept along at just five miles an hour as it wobbled along the Texas coast. In some flood-ravaged areas, the downpours continued unabated for up to four days.

A second important contributing factor is tied to higher ocean heat content. The ocean is the largest solar-energy collector on earth. More than 90% of the warming that has happened on Earth over the past fifty years has occurred in the ocean. The vast supply of heat stored in the oceans continues to rise every year as ocean temperatures warm, making tropical cyclones stronger, bigger, and longer lived. Another 2018 study led by Kevin Trenberth, an expert on climate change and its influence on hurricanes, concluded that Harvey's record rainfall was fueled by water temperatures in the Gulf of Mexico that were warmer than at any time on record, averaging over 86°F. Trenberth also noted that Harvey's large size meant that "even after landfall, its circulation extended well out over the Gulf, where a continual flow of moisture fed and prolonged the storm, long after most storms would have died." It has also been well documented that higher ocean heat content fuels more intense rainfall. In the case of Imelda, rainfall rates of up to 11" per hour were observed. The combined effects of prolific rainfall rates and sluggish movement led to Harvey dumping the largest volume of rainfall ever recorded in the United States.

TROPICAL TROUBLES IN NORTH TEXAS

Tropical systems do not often track into North Texas, primarily because hurricanes and tropical storms, after making landfall along the coast, typically recurve to the northeast as steering winds tend to take them through East Texas and then into Louisiana. When tropical systems do move into North Texas, as Tropical Storm Hermine did in 2010, they can have severe consequences. Hermine made landfall just south of Brownsville without much fanfare, but then tracked almost due north as it rode up the western side of a ridge of high pressure anchored to our east. The storm's track, just west of I-35, channeled a plume of deep tropical moisture into North Texas, feeding intense bands of thunderstorms that trained (move repeatedly over the same area) along the entire I-35 corridor. Up to 10" of rain fell in the DFW area in less than twenty-four hours. More than sixty water rescues were made in Johnson County alone after flash flooding swamped numerous homes. In Alvarado, one person was killed after he drove his car into a flooded street and was swept away. Arlington was also hit hard by the tropical deluge. One apartment building was overcome by up to 8' of water after Rush Creek flooded, forcing ninety people to evacuate. More than fifty homes were severely damaged. The city of Arlington was able to acquire and demolish forty-nine of the fifty homes and the apartment complex as part of a $17 million buyout to move people out of the flood-prone Rush Creek area.

NORTH TEXAS FLASH FLOOD STUDY REVEALS DISTURBING TREND

While tropical systems have produced several notable flood events over the decades, history shows that, in North Texas, the unnamed storms are responsible for a lion's share of the fatal encounters with floodwaters. Clusters or complexes of thunderstorms often tap into rich Gulf moisture and can produce rainfall rates that often exceed 3" per hour. In many cases, a stalled frontal boundary serves as the focus for development. The threat of flash flooding

Love Field Flooding *(KDFW-FOX 4 News)*

increases with training and slow-moving or stationary storms, especially if the ground is saturated from recent heavy rainfall. Similar to a wet sponge, waterlogged soil cannot soak up additional moisture, and much of the rain runs off into swollen rivers and creeks, causing them to flood. The FOX News video clip below shows what happened at a Love Field parking garage in the aftermath of severe flash flooding in April 2019.

A study conducted by meteorologists Ted Ryan, Lauren Tyler, and Stacie Hanes, from the National Weather Service office in Fort Worth, examined North Texas flooding events from 1995 through 2018. The most disturbing revelation in this study was the disproportionate number of young children who died as a result of flash flooding. Of the 104 flood-related fatalities in the twenty-three-year period studied, eighteen were children under ten years old, the most of any age-group category. Over 60% of these deaths occurred in vehicles, and in several cases the adult drivers drove around flood barricades with their very young passengers. The age group that ranged from twenty to twenty-nine accounted for seventeen deaths, and the group ranging from thirty to thirty-nine accounted for fifteen.

The study showed that the months of May and June had the highest number of flash-flood events. This is expected, as those are typically two of the stormiest months of the year and, in this period of record, they accounted for the highest monthly rainfall totals. The research also revealed three peak times when flash-flood events occurred. The highest peak time was from 7 p.m to 10 p.m., coinciding with a high volume of commuter traffic. A second peak occurred from 8 a.m. to 10 a.m., during the latter portion of the morning commute, and a third active period was noted between 1 a.m. and 3 a.m. While you might expect far fewer motorists on the road after midnight, those who do venture into floodwaters at that hour face the peril of darkness that prevents a driver from judging the depth of water and the condition of the underlying roadbed.

THE MOST DANGEROUS LOCATIONS FOR FLOODING

Based on the data compiled during the National Weather Service research study, including where fatalities, high-water rescues, and evacuations occurred, the authors of the study assembled a subjective ranking of the eleven most dangerous flash-flood locations in their forecast jurisdiction. The region encompassing these locations includes all of North Texas and extends as far south as the Killeen and Waco areas. They are as follows:

1. Nolan Creek Basin (Killeen-Harker Heights)
2. Turtle Creek (Dallas County)

3. Johnson Creek (Tarrant County)
4. Highway 67 Corridor (Keene-Venus area in Johnson County)
5. Waco (McLennan County)
6. Southwest Fort Worth
7. Duck Creek (Northeast Dallas County)
8. Mountain Creek (Comanche County)
9. Post Oak Creek (Grayson County)
10. Robinson Creek (Hood County)
11. Buffalo Creek (Johnson County)

While the majority of these locations are outside of the Metroplex, they all contain small creeks that easily flood neighboring parks, streets, and roads during heavy rainfall. At the top of the list is the Nolan Creek Basin, which features numerous bridge crossings. Since 1995, eight separate fatality events have occurred in this part of central Bell County along Highway 190 from the Fort Hood/Killeen area to Nolanville.

Four hot spots for flash flooding lie within Dallas and Tarrant counties. One is the Turtle Creek area, just a few miles north of downtown Dallas. This has been the site of two major flash-flood events (Mayfest 1995 and the March 2006 flood) in which rapidly rising waters swamped several vehicles, and Good Samaritan rescuers were sucked into whirlpools that developed above manholes. An SMU police officer became the latest fatality in 2016. He was working an off-duty security job guarding a home under construction when the fast-rising waters overtook his vehicle. His body was found seven weeks later in the Trinity River, about three miles from where he was pulled into the swollen Turtle Creek.

Johnson Creek in Arlington raises a particular concern due to its proximity to the University of Texas at Arlington, as well as several entertainment venues, including AT&T Stadium, Six Flags Over Texas, and Globe Life Park. In 2007, three children playing in the creek were carried away and nearly drowned, and in 2018, a twenty-three-year-old doctoral student drowned after he was swept off a bridge into the creek by raging floodwaters. The stretch of Duck Creek that runs through Garland is notoriously dangerous when torrential rain affects the area. Flooding of Duck Creek has been responsible for two fatalities, during two separate flood events, and at least three high-water rescues. Several parks and a golf course run adjacent to the creek bed, while several city streets cross the creek.

THE DEADLIEST FLASH FLOOD EVENT
IN NORTH TEXAS HISTORY

Mention the Mayfest storm in Fort Worth and locals instantly recall the huge hail that pummeled festivalgoers along the banks of the Trinity River. But it was the "second act" of the event that proved to be deadly. As a squall line overtook the hail-producing supercell, the merging complex of storms slowed to a crawl as it headed east into Dallas County. Torrential downpours with rainfall rates of up to 8" per hour turned roads and highways into rivers and triggered widespread flash flooding in the Metroplex. Drivers on the North Central Expressway were stranded under overpasses in water that rose to their windshields, while dozens of others had to be rescued from the rooftops of their vehicles on the streets leading out of downtown Dallas. One man died while trying to rescue three people caught in a flooded underpass when he got sucked into a storm drain. Just east of downtown at Music Hall in Fair Park, the unprecedented rain stopped a performance in the middle of the first act. More than one hundred vehicles had floated across the parking lot, stranding six hundred theatergoers, who had to be transported by city buses. At Baylor University Medical Center, patients were rushed by elevator to the hospital's upper floors after the emergency room flooded.

Tragically, fifteen people died in Dallas County and one person died in Tarrant County when he drove into a swollen creek. Many of the Dallas County victims had driven into floodwaters and were swept away, and other victims were children who drowned while playing near drainage culverts. Four other people died in Dallas County: two were struck by lightning and two died after a roof collapsed under the immense weight of standing water. In total, twenty people were killed, and seventeen of those deaths were attributed to flash flooding.

TURN AROUND DON'T DROWN

People often underestimate the power of water. It takes only 1.5' to 2' of rushing water to carry most vehicles off the roadway. The "Turn Around Don't Drown" slogan urges drivers to turn around and take a different route when they encounter water running across the roadway. In 2018, high water frequently swamped roads all across the Metroplex during the month of September, when a record-breaking 12.69" of rain fell. The cloudbursts brought flash flooding, which claimed the lives of four people in Tarrant County alone. Three people died on September 8 in two separate incidents less than a mile apart in East Fort Worth, when 5" of rain fell in less than three hours. A sixty-nine-year-old man drowned when his car was swept off a road that feeds into Lake Arlington. The other incident involved an eighteen-year-old woman and her two-year-old daughter. Their car stalled in high water on the Loop 820 Access Road. As other vehicles passed by, the wakes they created pushed her car into a drainage culvert, where it sank in 15' of water.

In the aftermath of the flash-flooding deaths, researchers associated with the CASA weather radar project in North Texas conducted a survey in the hope of shedding some light on the rash of deadly incidents. Their survey drew responses from more than one thousand people in the area, 226 of whom had encountered flooded roads during the first two weeks of September. Of the 226 people who had come across a flooded road, 61% continued driving through the water while only 39% turned around. When asked why they ventured into the water, 70% of those who crossed said they perceived the water level to be manageable by their vehicle, and 50% said they made the decision after watching other motorists drive through. Perhaps even more concerning, deep water did not deter a fair number of drivers. When they estimated the water to be greater than 2' deep, 38% of the respondents still proceeded through the floodwaters.

While preventing drivers from entering flooded roadways is difficult, several cities in North Texas have taken measures to do so. Fort Worth officials have identified 280 locations as being at risk of flooding. At fifty-two of the city's highest-risk sites, flood gauges have been installed to measure water levels as part of the High Water Warning System (HWWS). Each low-water crossing has roadside flashers that are triggered to warn drivers of a flood hazard. At the same time, text and email alerts are sent to first responders. The city of Dallas has forty flooded-roadway warning signs. Each sensor monitors the elevation of a nearby stream and reports changes every three minutes to the central computer. When floodwater reaches the edge of the roadway, a float switch tells the sensor to signal the sign to change to the warning text and turn on the flashing lights. Emails are sent to staff and the appropriate street-services district, alerting them of the need to place barricades at the location as soon as possible. In the alarm state, the lights alternate flashing and the sign changes to "Do Not Enter High Water." The city of McKinney, in collaboration with the North Central Texas Council of Governments and the Texas Water Development Board, is working to install its own early flood warning system at five points along two roads near Wilson Creek, which is highly susceptible to flash flooding. These areas include Valley Creek Trail, Park View Avenue, and US 75 near Towne Lake Park. The system will monitor water depth and, when the water is high, will activate flashing lights to warn drivers of flooding. Like the systems in Dallas and Fort Worth, city staff will be alerted via email when the flashers are set off.

THINKING OUTSIDE THE BOX TO REDUCE STORMWATER RUNOFF

Slow it down, spread it out, and soak it in are the three core principles of the Integrated Stormwater Management (ISWM) program. ISWM is a regional initiative administered by the North Central Texas Council of

Governments (NCTCOG), which brings together representatives from cities and counties to find collaborative solutions to regional environmental challenges. Currently, eighteen cities in Dallas, Tarrant, Collin, and Denton counties participate in the program. ISWM encourages creating natural areas and streams to treat and reduce the flooding impact of stormwater runoff from streets and parking lots, and the program's influence is evident in several projects throughout the Metroplex. The Fort Worth Nature Center and Refuge built a new "green parking lot" using recycled crushed concrete and compacted gravel combined with five bioswales (landscaped drainage swales) to allow stormwater to soak in rather than run off. A downtown parking lot in Carrollton uses special Grasstone pavers that allow grass to grow on the pavers' interior to permit stormwater to infiltrate the pavement subsurface. The Green at College Park, a focal point along the city of Arlington's Center Street Pedestrian Trail, included recycled paving materials, native grasses and adaptive plants, and a dry creek bed to help manage rainwater and stormwater runoff that drains into Johnson Creek. ISWM members hope that by creating systems that mimic the natural environment, development can be integrated within a space in a way that is visually pleasing and does not increase runoff.

Stormwater Management Project *(NCTCOG)*

Mark England heard the phrase "it's just stuff" frequently in the days after his family lost everything they owned to the flood. While Mark thanks God every day that he and his family got out of their home alive, he admits that some of the material things they lost had enormous sentimental value: "Cribs that held our ten foster babies. Our dining table of twenty years where our homeschooling, years of family meals, holiday family reunions, and church potlucks had taken place. There's really no words to describe the shock your heart goes through." Mark had recently lost his job, and then the flood took their home, their cars, and all of their possessions. Joy said they couldn't have gotten through it without the outpouring of help and generosity from family, friends, and complete strangers. "People we didn't know brought us meals. People showed up with toys. The school brought us clothing. We were so surrounded by love from strangers." In fact, a couple they didn't know generously donated two RVs that enabled the Englands to move back to their Red Oak property. Despite the outpouring of support, the seven months following the flood were a daily struggle for Mark, Joy, and their five children, but their prayers were finally answered when Mark was offered a position as an associate pastor at a church in Wellington, Colorado, where they have settled and built a new home.

CHAPTER 9

HEAT WAVES, DROUGHTS, AND THE WATER SUPPLY

One of the biggest adjustments I had to make when I took the chief meteorologist position at FOX 4 in early 2003 was getting accustomed to the unrelenting Texas summer heat. June of my first summer was not too bad. Our swimming pool was refreshing, and the air conditioning kept our home a cool 72°F. To top it off, I was getting paid to tell people it would be sunny and hot yet again tomorrow. What a gig! Then July brought a barrage of triple digits. It was 105°F in the shade at three in the afternoon, but at least (I thought) the temperature would drop after sunset. Wrong! I had long heard of the urban heat island effect and now I was fully immersed in its suffocating splendor. The temperature was still 97°F at nine in the evening. By the end of July, it wasn't just the heat taking my breath away. The first glance at my utility bill nearly made me pass out. I soon learned that 78°F on the thermostat feels just fine. By

August, I had burned the back of my thighs a half dozen times sitting on my car seat, my prized green lawn was scorched, and that refreshing swimming pool was now a giant hot tub, with the pool thermometer registering a balmy 93°F. At long last, Labor Day weekend and the unofficial end of summer arrived. To my utter disappointment, however, I discovered I had to labor through at least one more month of sweat and sizzle. Hot weather in the summer in North Texas is a part of life, just as frigid winters are in Minnesota. Most people never get used to the extreme heat; they just find ways to cope with it.

STORMY SPRING CAN DELAY SUMMER SWELTER

You've probably heard the old sports cliché about a great shooter in basketball or a perennial All-Pro running back: "You can't stop him; you can only hope to contain him." The same sentiment applies to North Texas summers. Sometimes, our best hope of delaying the onslaught of summer heat is a rainy May that carries over into early June. Climatologically speaking, May is the wettest month of the year, averaging over 5" of rainfall. The stormy pattern occasionally continues into the first weeks of June with "northwest flow" storms that roll into North Texas well after midnight. These storm complexes form in the High Plains of Southeast Colorado and the Oklahoma and Texas Panhandles in the evening and are driven southeast by northwest winds in the upper atmosphere. They tend to hold together well after sunset due to a phenomenon called the nocturnal low-level jet: a belt of speedy, southerly winds roughly 1,000' to 3,000' above the ground that often develops during the late evening. Blowing at average speeds of thirty to fifty miles an hour, the nocturnal low-level jet can quickly transport Gulf moisture northward to sustain these thunderstorm complexes through the night.

As the northwest flow events wane in mid-June and the strong rays of early summer reign supreme, the effects of a wet, late spring can take a bite out of the early summer heat through increased evapotranspiration. Evapotranspiration is the combination of two simultaneous processes:

evaporation and transpiration, both of which release moisture into the air. During evaporation, water is converted from liquid to vapor as it is evaporated from wet soils. Through transpiration, water that was drawn up through the soil by the roots evaporates from the leaves of the lush, green vegetation. As moisture evaporates, the nearby air is cooled, which counters, at least to some degree, the warming from the sun. This may partly explain why the average date of the first 100°F day at DFW Airport is July 1. The slightly cooler temperatures, however, are offset by higher humidity levels. In fact, an examination of annual relative humidity reveals that it peaks around the Fourth of July and then slowly and steadily declines through the rest of summer as rainfall dwindles.

HEAT INDEX: HOW HOT IT REALLY FEELS

Early summer in North Texas can be steamy, owing to the cloudbursts we see on a regular basis from May through the middle of June. The ratio of how much moisture is in the air compared to how much the air can hold is called relative humidity (RH). Relative humidity can be misleading because, at warmer temperatures, the air can hold much higher quantities of moisture: roughly 7% more per 1.8° of warming. There is substantially more moisture in the air on a 95°F day with 50% relative humidity than on a 75°F day with the same 50% relative humidity reading. A better way to assess how much moisture is in the air is looking at the dew point. The dew point is the temperature to which the air must be cooled in order to reach saturation. The higher the dew point, the more moisture there is in the air. Very high levels of moisture in the air can make the outside temperature feel several degrees warmer, because the body cools itself by sweating, and sweat in turn evaporates and carries heat away from the skin. However, when the RH or dew point is high, the evaporation rate can be significantly reduced and the cooling process compromised. As a result, the body feels warmer. The heat index combines the air temperature with the relative humidity or dew point to determine what temperature the air actually feels to our human bodies. Below is a heat index chart, which shows the "feels like" temperature for a given temperature and relative humidity. For example, when the air temperature is 96°F and the relative humidity is 30%, the resulting heat index is also 96°F because your body is cooling itself efficiently in the "drier heat." Raise the relative humidity to 50% with the same 96°F temperature and the heat index becomes a stifling 108°F. When heat index values rise above 105°F, prolonged exposure and activity can significantly increase your chances of heat stroke.

Heat Index ("Feels Like Temperature")
Temperatures (°F)

Relative Humidity (%)	88	90	92	94	96	98	100	102	104
60	95	100	105	110	116	123	129	137	145
50	91	95	99	103	108	113	118	124	131
40	88	91	94	97	101	105	109	114	119
30	86	88	90	93	96	99	102	106	110
20	85	86	88	90	93	95	97	100	103

Heat Index Chart

THE SEARING SUMMER OF 1980

Back in the summer of 1980, there were two big questions on the minds of North Texans: "Who shot JR?" and "When will this dreadfully hot summer ever end?" Fans of the popular television show *Dallas* would have to wait until the "Who Done It" episode in late November to find out who pulled the trigger. Fortunately, the record-setting heat subsided long before the big reveal on *Dallas*. For decades, the "measuring stick" for North Texas summers was 1980, when twenty-nine daily high-temperature records were either tied or broken and the all-time record high of 113°F was established and then tied the following day. My friend and colleague, FOX 4 News anchor Clarice Tinsley, who came to KDFW from Milwaukee in 1978, recalls the summer of 1980 as brutal and unrelenting: "I remember that I didn't want to go outside to get the mail, much less work in [the heat], drive in it, or be in it." The deadly heat wave, which claimed the lives of thirty people in Dallas and Tarrant counties alone, dominated the newscasts. "That summer, the drumbeat of extreme heat was 'the news,' putting Dallas and Fort Worth in the national and international spotlight, with crews that came here to report on the impact of the punishing heat and how we all tried to cope in it."

A thick canopy of cirrus and altocumulus clouds associated with Hurricane Allen, spiraling over North Texas, blocked out enough sunshine to finally end the heat wave on August 4, after six consecutive weeks at or above 100°F. The 1980 records were so numerous and the heat so intense that meteorologists thought there would never be a summer to rival it. Then came the summer of 2011, which was punctuated by two long hot streaks. The first stretch of searing triple-digit heat began on July 2 and continued uwavering through August 10. After a mini four-day reprieve, North Texas sweltered through a twenty-day streak of temperatures that hit or exceeded 100°F lasting into Labor Day weekend. The hot, dry weather created dangerous fire conditions. Nearly twenty thousand wildfires burned four million acres, an area roughly the size of Connecticut.

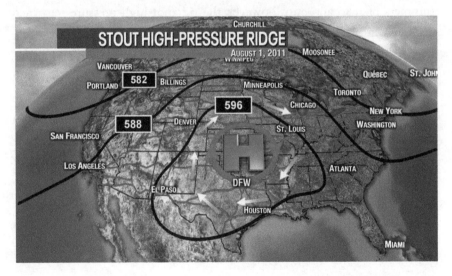

The 500 mb chart above depicts conditions in the middle troposphere, the approximate halfway point of the atmosphere, on August 1, 2011. The solid lines represent the level at which the air pressure reaches 500 mb, or about half the normal sea-level pressure of 1,013 mb. The 500 mb pressure level generally ranges in elevation from 4,900 meters (16,000') to 6,000 meters (20,000'). The "596" line on the chart indicates that the 500 mb pressure level is located at a height of 5,960 meters above sea level. The height of the 500 mb pressure level is related to the temperature of the atmosphere; the higher the temperature, the higher the height of the 500 mb level. Air expands as it is warmed, and if the vertical column of air is warmed, the column expands upward. When the height contours bend to the north, this is called a "ridge." Strong ridges are usually associated with warm, dry weather beneath them. The wind arrows denote the clockwise circulation around the powerhouse ridge, centered over the Kansas/Oklahoma border, on what turned out to be the hottest day that year, with a scorching high of 110°F at DFW Airport.

During most summers, we typically see the ridge of high pressure temporarily break down or retrograde into the Desert Southwest. This opens the door for weak cold frontal passages and increases the chances of afternoon thunderstorm activity. More importantly, the heat eases slightly as temperatures fall back into the nineties and occasionally the eighties. In the summers of 1980 and 2011, however, the ridge of high pressure remained anchored in place for several weeks, enabling the North Texas "blast furnace" to operate 24/7. Recall that high pressure aloft causes the air to sink, which promotes clear skies and allows the sun to heat the air efficiently. This downward motion also compresses and warms the air in the lower atmosphere while simultaneously trapping heat that attempts to rise from the earth's surface. Any thunderstorms that do manage to fire up over the High Plains to our west and northwest are diverted around the bubble of high pressure, keeping North Texas high and dry.

SUMMER SHOWDOWN: 1980 VERSUS 2011

The graphics below compare 1980 and 2011 in six different statistical categories. The summer of 1980 claims the all-time record high temperature of 113°F, the highest average high temperature of 101.6°F, and the longest stretch of 100°F days. The summer of 2011 recorded the most 100°F days (seventy-one of them!), the highest average temperature of 90.5°F, and the highest average low temperature of 79.8°F. A closer examination of the numbers reveals that 1980 had more extreme heat. The mercury soared above 105°F on twenty-eight days in 1980 and 110°F or higher on five days, compared to only fifteen times and one time respectively in 2011. So based on those numbers, you would have to give the nod to 1980 as the hottest summer. But let's not discount temperatures after dark. The average low of

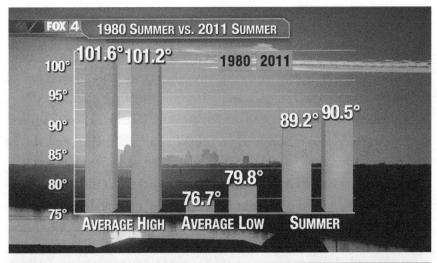

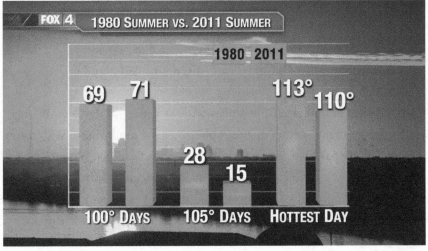

almost 80°F in 2011 was more than three degrees warmer than in 1980, and residents got no relief from the heat at night. Additionally, this implies that humidity levels were higher in 2011. From that perspective, you could argue that 2011 was even more unbearable. So, what's the verdict in this show-down of the two "summer heavyweights?" Based on the hottest three-day stretch in DFW history, with consecutive highs of 113°F, 113°F, and 112°F, it's hard not to give the slight edge to 1980.

TEXAS HEAT BEGETS DROUGHT

Isaac Cline, the chief meteorologist for the Galveston, Texas office of the US Weather Bureau during the Galveston hurricane of 1900, once described Texas as "a land of eternal drought occasionally interrupted by biblical floods." Indeed, Texans are familiar with the "all or nothing" cycle of drought and flood. According to the Texas Water Resources Institute, there have been more than a dozen major droughts since the 1930s and almost all of them have ended with cloudbursts that led to widespread flooding. Some droughts lasted for several months while others continued for years. A prime example of the "feast or famine" nature of rainfall played out in 2018. After delivering a whopping 11" of rain in February, the skies barely sprinkled for the next six months, during which the DFW area measured a paltry 9" of rainfall. Fall of 2018 flipped the script from drought to deluge with a record-breaking 25" of rainfall in September and October. While the rain brought an end to water restrictions and filled up area lakes, the fast and furious downpours led to several episodes of flash flooding.

Drought is a complicated phenomenon that generally develops in stages and impacts many sectors of the economy. As a result, those who moni-tor drought separate it into four different classifications: (1) meteorolog-ical drought, (2) hydrological drought, (3) agricultural drought, and (4) socioeconomic drought. Meteorological drought is directly related to the departure of precipitation from the norm, but due to climatic differences, conditions that might be considered a drought in Florida may not be a drought in Texas. Hydrological drought occurs when surface and subsurface water supplies drop below normal. Agricultural drought takes place when crops are adversely affected due to insufficient soil moisture. Finally, socioeconomic drought ensues when the demand for various commodities exceeds the supply as a result of drought conditions.

Meteorological drought can begin and end rapidly, while hydrological drought takes much longer to develop and recover from.

MONITORING DROUGHT

Instituted in 2000, the us Drought Monitor is a weekly updated index that measures drought in these various sectors. Drought severity is measured on a scale ranging from D0, indicating abnormally dry conditions, to D4, indicating exceptional drought conditions. The three intermediate designations are D1 (moderate drought), D2 (severe drought), and D3 (extreme drought). Since the index's inception in 2000, the longest-duration drought in Texas lasted from May 2010 to July 2015, with conditions across the state that ranged from D1 to D4. The drought peaked during the week of October 4, 2011, when a remarkable 88% of Texas experienced D4 drought conditions. The five-year drought ended with record-breaking rainfall in May 2015, with measurements totaling 16.96" at DFW Airport. North Texas lakes that had been draining like bathtubs leaving lake bottoms exposed, were now overflowing. On March 1, 2015, the water level at Lake Grapevine was 524' above sea level, or roughly 11' below normal. By May 31, the water level had risen 38" to an elevation of 562' above sea level, which was a staggering 27' above normal.

The drought of 2000, while much shorter than the drought of the 1950s, produced the longest rain-free period in recorded history for the DFW area. No measurable rain fell for eighty-four consecutive days, from July 1 through September 22. Other severe droughts occurred in 2006 to early 2007, and again in 2008 through 2009. The Drought Monitor clearly illustrates how frequently these parched periods of weather recur and how quickly drought conditions become established and intensify.

HISTORIC DROUGHTS IN TEXAS

The "Dust Bowl" of the 1930s, which affected a large portion of the Plains, from southern Nebraska to West Texas, was a result of severe drought conditions and poor farming practices. Farmers had deep-plowed millions of acres of land during the previous decade, removing the established native grasses that helped trap soil and moisture. The drought led to widespread crop failure, and this left the unanchored soil exposed to the intense storms of spring. The storm that raged on April 14, 1935, known as "Black Sunday," lifted three hundred thousand tons of topsoil into the air, turning daylight to darkness from Liberal, Kansas, to Amarillo, Texas. Reporters covering the storm nicknamed the region "The Dust Bowl." These "black blizzards" carried dust as far north as Chicago, Illinois, and as far east as New York City, New York, and Washington, DC. In North Texas, the air was often filled with choking billows of dust.

Millie Stuck was born on a farm in 1915 in Mills County, Texas, about a two-hour drive southwest of Fort Worth. During a conversation over lunch in 2019, Millie told me there were many days during those Dust Bowl years that were pure misery. "The skies turned black and you could hardly breathe. We had no air conditioning back in those days, so the windows and doors were open, and the dust and dirt would blow in. We swept up enough dust to fill up two large buckets every day."

The Dust Bowl turned out to be a mere prelude to a drought of epic proportions. The seven-year drought of the 1950s was the worst on record in the state of Texas. Some towns went completely dry, and water had to be brought in by truck or rail. Farmers saw their crops fail on a massive scale, and overgrazed pastures and skyrocketing prices for feed forced cattlemen to sell off their herds. The "Drought of Record" transformed Texas from a rural to an urban state. From 1950 to 1960, Texas lost nearly one hundred thousand farms and ranches and close to 10% of its rural population. The multiyear drought also hit big cities hard. Officials in Dallas had to pump water in from the Red River to meet the city's water demands. The high concentration of salt in the brackish water threatened patients with compromised kidney function, choked landscape plants, and corroded car radiators.

1930s Texas Dust Storm (*NOAA–George E. Marsh Album*)

According to historian Thomas Hatfield, the searing heat threatened to burn the Cotton Bowl turf to a crisp, forcing groundskeepers to drill a water well in the end zone. "The greatest outrage to City Hall was the Cotton Bowl, that emporium of gladiator pride, having to drill its own well within the stadium in order to water the turf because the Dallas waterworks could not furnish the means for such lavish irrigation…Good strike made at a depth of only 35' just twenty-seven yards south of the goal posts." All but ten of the state's 254 counties were declared drought disaster areas by President Eisenhower.

Finally, the skies opened up in the spring of 1957, unfortunately in spectacular fashion, with several extreme rainfall events. One of the largest, dubbed

"The Day of the Big Cloud," occurred on April 24, 1957. The storms, accompanied by damaging hail and multiple tornadoes, dumped 10" of rain over a large portion of Texas in less than six hours. The spring floods killed twenty-two people and forced thousands to evacuate their homes.

DROUGHT-FUELED WILDFIRES

In 2011, Texas experienced the worst single-year drought in recorded history. The lack of rain contributed to an unprecedented wildfire season with some of the largest, most destructive blazes on record. Ironically, a parade of four tropical systems in Texas in the summer of 2010 had set the stage for the fires. The tropical deluges stimulated rapid grass growth across the Texas Plains. The vegetation served as kindling once it was cured by the first winter freeze. More than four million acres were scorched and 2,947 homes destroyed by 31,453 wildfires that were ignited statewide. Firefighters fought valiantly and saved nearly thirty-nine thousand homes—more than thirteen times the number of those destroyed. Sadly, four firefighters gave their lives fighting the flames and six civilians perished.

Possum Kingdom Lake Fire *(Brent Isom Photography)*

One of the largest fires was a grouping of four blazes in Stephens, Young, and Palo Pinto counties known as the "Possum Kingdom Complex" fire. The monstrous wildfire swept into Possum Kingdom State Park in April 2011. The rugged terrain around the lake combined with fifty-mile-an-hour wind gusts to create a "firestorm." Flames soared to heights of up to 100' as the fires raced uphill and leapt from treetop to treetop.

Ronnie Ranft, a thirty-year veteran firefighter and chief of the Possum Kingdom Lake Volunteer Fire Department for sixteen years, said it was the most extreme fire behavior he had ever witnessed. "I had never seen anything act so angry. We would see three, four, five wind shifts every day." The area surrounding Possum Kingdom Lake was covered with mountain cedar (juniper), which is considered a high-risk fuel because it burns so intensely. "You could feel the heat a mile away," said Ranft. The former fire chief reported that, while they lost a great number of homes, many individuals' valiant efforts helped save others from burning. "One hero was a game warden. He chased down a Sea-Doo watercraft that had caught fire (when flames reached the dock it [had been] moored to) and was drifting toward the peninsula. Had it gotten to shore, it would have caught the grass and trees on fire and set the whole peninsula on fire." The Possum Kingdom Complex fire scorched nearly 127,000 acres and consumed 167 homes. On a brighter note, Ranft says the outpouring of support blew him away. "We had locals cooking food for us and bringing food in. One local store owner gave us the keys to his store and told us to get whatever we needed."

Possum Kingdom
Lake Fire Video
(Mike Warner)

The Possum Kingdom Complex fire was not the worst fire of 2011. On Labor Day weekend, a series of deadly wildfires broke out in Bastrop State Park, about thirty miles east of Austin, and soon dwarfed the Possum Kingdom fire. Fanned by gusty outflow winds from Tropical Storm Lee, the fire grew quickly and spread into nearby communities, prompting mass evacuations. Fed by tinder-dry vegetation, the Bastrop County Complex fire burned for nearly a month and resulted in four fatalities. When it was finally contained in early October, it had burned more than thirty-four thousand acres and 1,645 homes, making it the most destructive fire in Texas history. The 2011 fire season in Texas was a wakeup call demonstrating the mounting threat that wildfires pose as cities and towns expand into previously rural areas. In Texas, 80% of wildfires occur within two miles of a community. Given that it takes less than thirty minutes for fire to travel two miles, this leaves residents little time to evacuate. These rural communities also face longer response times, due to their greater distance from emergency services.

PLANNING TO MEET THE FUTURE WATER DEMAND

Two things appear certain over the next fifty years: there will be an explosion in population growth in Texas and there will be many more droughts to endure. Texas' population is expected to soar from 29.5 million in 2020 to 51.5 million in 2070. This is projected to result in a 17% increase in water demand. In the event of a repeat of the 1950s drought, Texas' water woes could spiral into a water crisis. The total water shortage could reach 8.9 million acre-feet in 2070 (one acre-foot is equal to a sheet of water covering one acre of land to a depth of one foot).

The 2017 Texas State Water Plan (the next one will be drafted in 2022) recommends 5,500 water-management strategies to meet future demand. These strategies include conservation, the construction of new reservoirs, water reuse, seawater and groundwater desalination, and groundwater wells. If implemented, these measures would provide 8.5 million acre-feet of additional water per year by 2070, but at an estimated cost of $63 billion. If the strategies are not undertaken, by 2070, approximately one-third of the population will have less than half of the municipal water supplies they need.

A major component of the water plan is the construction of twenty-six new reservoirs. Tough environmental regulations, contentious disputes with landowners, and spiraling costs, will make getting all of the proposed reservoirs built a monumental task. One of those new reservoirs is already under construction near Bonham in Fannin County. Bois d'Arc Lake is the first major reservoir constructed in Texas since 1999 and the first to be built in the Dallas-Fort Worth region since Joe Pool Lake in 1986. The project, which began in 2018, will create a 16,641-acre lake that will supply water for 1.7 million people in eighty communities served by the North Texas Municipal Water District. Funding for the $1.6 billion reservoir will come from the Texas Water Development Board's State Water Implementation Fund for Texas.

Bois d'Arc Lake Construction (*NTMWD*)

Even if all twenty-six new reservoirs are constructed, more than two million acre-feet of water will be needed to meet the municipal water demands in the year 2070. State Climatologist John Nielsen-Gammon insists the only way to meet our future water needs will be through conservation measures: "People need to be aware how rapidly drought can develop and how bad it can get. We can't be complacent. We must be very responsive to requests for conserving water and restricting water use when a drought is developing." Texans appear to be heeding the call for conservation. Just one year after unveiling their 2017 plan, the Texas Water Development Board reduced their water demand projections from 21.6 million to 19.2 million acre-feet per year based on lower-than-expected water usage.

Organizations like Texas Water Trade and the Texas Living Waters Project (TLWP), nonprofit groups based in Austin, believe meeting our future water needs will require nontraditional methods. They contend that the state's historic reliance on groundwater, rivers, and reservoirs is not a feasible long-term solution. These resources have become overburdened due to increased demand and lose tremendous amounts of water due to the exposure to wind and high temperatures, the drivers of evaporation. In the extreme drought year of 2011, North Texas lakes lost an average of 7' of water due to evaporation.

These conservation groups believe the state needs to provide incentives for using green technologies to harvest rainwater, and recycle wastewater and stormwater to bolster local water supplies. These forward-thinking strategies are already paying dividends. Toyota Motor North America's headquarters campus in Plano uses a state-of-the-art rainwater capture system to collect up to four hundred thousand gallons of water for irrigation, estimated to save more than eleven million gallons of drinking water annually. Austin's Central Library uses rainwater and condensate from its air conditioning systems to irrigate its landscaping and to flush its toilets. The TLWP contends that many more need to jump on the conservation bandwagon. They claim there are many businesses and communities throughout the state that are doing relatively little to reduce water consumption. The alliance believes that more participation in cost-effective conservation will reduce the need for expensive, new water-supply infrastructure and close the gap between available water supplies and future water demands.

CHAPTER 10

WINTER WEATHER

ARCTIC BLASTS TO ICEMAGEDDON

The mere mention of snow or ice in North Texas generates quite a buzz. Some of the most frequently asked questions we receive from FOX 4 viewers have to do with snow and ice: Will we see a white Christmas this year? Will we get enough snow for school to be cancelled tomorrow? How bad will the roads be in the morning? Sadly, snow lovers and "snow-day lovers" (there's a difference) in North Texas rarely hear answers that make them happy. Computer models often hint at snow days in advance, but the hopes of those who would love to build a snowman or go sledding at Flagpole Hill are almost always dashed.

For snow to occur in North Texas, all the right pieces have to fall perfectly into place. The margin between rain and snow or rain and ice in North Texas is razor thin and may boil down to a one- or two-degree difference in temperature or a fifty-mile shift in the track of the storm. The timing of cold fronts is also critical and often presents cases where "cold air is chasing the moisture." More often than not, by the time temperatures are cold enough for snow, the moisture has been swept off to the east ahead of the cold front. Because one ingredient is usually absent, North Texas typically gets cold rain. The average annual snowfall for the Dallas-Fort Worth area is a meager 1.7". Cities northwest of Fort Worth such as Bowie and Jacksboro, which sit at higher elevations, average closer to 3" annually. Yet these are only averages; we may go several winters with only a few stray flurries and then get hit with heavy, consecutive snowfall events. In the thirty years from 1989 to 2019, DFW has seen a grand total of 51" of snowfall, which is how we arrive at our yearly average of 1.7", although distribution is remarkably uneven. A closer look at the data reveals that 23" of this snow fell in just two seasons, including a whopping 17.1" during the winter of 2009. Years with very little accumulation are the standard. In twelve of the thirty winters mentioned above, a trace or less was measured at DFW Airport.

IT USUALLY BEGINS AS SNOW

Even in snowless winters, North Texas usually sees one or two ice events, which can create commuter chaos. A dusting of snow causes very little disruption, but a thin coating of ice can wreak havoc. An ice event can come in the form of either freezing rain or sleet or both. Sometimes, one storm can produce a potpourri of all four major winter precipitation types: snow, sleet, freezing rain, and rain. Let's take a look at how each of these forms.

In winter, precipitation usually begins as snow falling through the upper portion of the clouds, generally 10,000' and higher, where temperatures are below freezing. If the layer of air from the cloud base to the ground is at or below freezing, the precipitation remains in the form of snow. There are times when surface temperatures can be in the 35°F to 40°F and snow will be falling. Why doesn't the snow melt at these temperatures? It actually does, to an extent. In order for snowflakes to survive their trip to the surface with temperatures several degrees above freezing, the air they are falling through must not be saturated, and the wet-bulb temperature must be at or below freezing. The wet-bulb temperature is the lowest temperature to which air can be cooled by evaporating water into it. As the snowflakes begin to melt, the melting creates evaporative cooling, which cools the air surrounding the snowflake and slows the melting process. A rule of thumb I use to determine the potential for evaporative cooling is the "one-third rule." The amount of "wet-bulb cooling" is roughly equal to one-third of the difference between the air temperature and dew point, referred to as the "dew point depression." For instance, if the surface temperature is 38°F and the dew point temperature is 20°F, the dew point depression is 18°F. Divide 18°F by three and you get 6°F cooling. So, in this case, evaporative cooling will take the temperature down 6°F from 38°F to 32°F, cold enough for the snow to not melt.

ADDING WARM AIR CAN CREATE AN ICY MIX

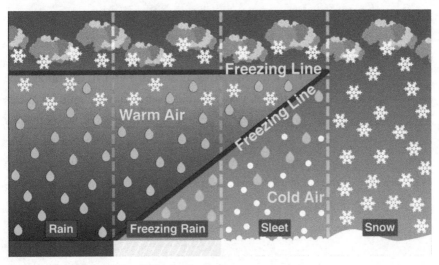

Precipitation Types *(National Weather Service)*

In North Texas, it is common for mild air to be blown in by south to southwesterly winds above the surface, forming a warm layer, often referred to as a "warm nose," where temperatures are above freezing. If the warm nose is shallow, the snow will partially melt as it falls through the layer. These slushy drops refreeze as they encounter a deep layer of freezing air above the surface and reach the ground as ice pellets called sleet that bounce on impact. Freezing rain occurs when snowflakes descend into a deeper layer of warm air and melt completely. If surface temperatures are at or below freezing, the raindrops can freeze upon contact, creating a glaze of ice on the ground, trees, power lines, and other objects. A significant accumulation of freezing rain lasting several hours or more is called an ice storm. An ice accumulation of just half an inch can add five hundred pounds of additional weight to one power line. Power lines are designed to handle up to half an inch of ice and forty-mile-an-hour winds. Heavier accumulations, combined with strong, gusty winds, can lead to widespread power outages.

Freezing rain accumulations are highly dependent on surface temperatures and can be impacted by the intensity of the falling rain. If surface temperatures are hovering around the freezing mark, minor ice accumulations are generally confined to elevated road surfaces like bridges and overpasses that are exposed to cold air from above and below. If surface temperatures are below 30°F, secondary and primary road surfaces become increasingly susceptible to ice accumulation, especially if temperatures remain below freezing for several hours. There are times when heavier rain can discourage ice accumulations. The heavier rain may occur as a surge of very mild air glides over much colder air hovering near the ground. In these cases, temperatures aloft (around 4,000' to 6,000' above the surface) may be in the forties. The heavier rain can drag some of this warm air to the surface. Additionally, the warmer raindrops tend to resist freezing even if temperatures at the surface are hovering around 32°F.

THE POLAR VORTEX AND BIG ARCTIC OUTBREAKS

The polar vortex media frenzy in recent years has generated a lot of hype and misunderstanding regarding this atmospheric feature. Contrary to what is often written or implied by news headlines, it is not a new phenomenon at all; it has always existed. The polar vortex is just a large, swirling mass of bitterly cold air located above the poles at an altitude of roughly 30,000' in the polar stratosphere. Recall that the troposphere is the layer of the atmosphere closest to Earth, where nearly all weather takes place. The stratosphere is the next layer above the troposphere and is the home of the ozone layer. The polar vortex weakens in summer and strengthens in winter as the polar stratosphere is plunged into complete darkness. Receiving no sunlight at all in winter, temperatures in the polar stratosphere drop to −100°F or colder. The term "vortex" refers to the counterclockwise jet stream, called the polar night jet, that forms to balance the huge difference between the frigid air near the poles and warmer air at lower latitudes. The polar night jet flows in a complete circle around the pole and typically keeps the frigid air corralled over and near the pole, though occasionally, this jet stream flow can be disrupted and become weaker. The polar vortex is similar to a spinning top. When it is spinning fast, that force and momentum keep the Arctic air bottled up near the poles. But when the vortex is weaker and spinning more slowly, it wobbles and becomes unstable, and the vortex can be displaced or split and can send lobes of frigid air southward into the middle latitudes.

NASA's Atmospheric Infrared Sounder (AIRS), aboard the *Aqua* spacecraft, captured the movement of the polar vortex that led to multiple invasions of Arctic air during the winter of 2013. The first polar plunge sent the temperature at DFW Airport spiraling downward from near 80°F on December 4, 2013, to a high of 26°F on December 7, 2013. The icy chill hung around for the better part of a week before we thawed. Another Arctic blast arrived shortly after the New Year. On January 4, 2014, we sat at a balmy 70°F during the afternoon before cratering to 15°F by sunrise on January 6, 2014. Watch the event unfold in the video link.

Polar Vortex animation *(NASA Jet Propulsion Lab)*

DECEMBER 1983 COLD WAVE

Many intense Arctic outbreaks are tied to episodes in which the polar vortex breaks down, sending lobes of bitterly cold air southward. This was the case in December 1983, when much of the central and eastern United States experienced bone-chilling temperatures, including the coldest holiday season in decades. More than 125 daily low-temperature records were set across the country on Christmas Day. The severe cold spell reached all the way into northern Mexico, with several freezes into the Lower Rio Grande Valley and the Gulf Coast, where Galveston plummeted to 14°F. The December 1983 deep freeze bottomed out with a low of 5°F at DFW Airport on the morning of December 22, a temperature that is still far from the lowest ever recorded in North Texas. That record was set on February 12, 1899, when the mercury plunged to −8°F. The last official sub-zero reading in the DFW area took place on December 23, 1989, with a low of −1°F. What the 1983 Arctic outbreak lacked in sheer intensity, it more than made up for in staying power. This cold wave was remarkable for its longevity. Once the mercury dipped below 32°F at 7 a.m. on December 18, it remained below freezing for the next 295 hours, until 2 p.m. on December 30. The seemingly endless polar plunge broke eleven daily temperature records and established a new monthly record. With an average temperature of 34.8°F, December of 1983 was 12.1°F below average, and easily rewrote the record books as the coldest December in history, dating back to 1899.

Blocking Pattern, December 1983

Like many Great Plains Arctic outbreaks, the December 1983 event featured a stout blocking ridge of high pressure over the North Pacific and Alaska. This blocking high acts like a boulder in a stream, forcing weather systems to move around it. These blocks can remain in place for several days or even weeks, causing the weather pattern to stagnate. To the east of the ridge, a north-to-south flow gave cold air from the Arctic Circle a direct pipeline southward into the United States. Because the blocking high remained anchored in place, it allowed repeated reinforcing blasts of Arctic air to invade the United States. The highly amplified jet stream pattern likely led waves in the atmosphere to break down the polar vortex. As a result, the polar vortex shifted southward, with smaller gyres splitting off and migrating into the lower forty-eight states. As Arctic air poured south of the border, it created one of the most expansive surface high-pressure areas ever observed across North America. DFW set its all-time highest barometric pressure reading of 31.06" of mercury on December 24. Not surprisingly, with the dome of Arctic high pressure at its peak, on Christmas Eve the high temperature at DFW Airport only reached 13°F, the second-coldest high temperature ever. Eagle Mountain Lake and Lake Worth completely froze over. Day after day of bitterly cold weather took its toll. Plumbers were on call 24/7 attending to a flood of emergency calls for broken pipes. Agricultural losses and damage to landscaping were staggering; the damage to agriculture in North Texas alone was estimated at $50 million. Thirteen people statewide, six of them in North Texas, perished due to the cold.

NORTH TEXAS NUANCES

Do you ever wonder why areas north and west of the Metroplex seem to get more ice and snow, while areas south and east get more rain? There are a couple reasons for this. The first has to do with elevation. Allow me to put a twist on the old Jimmy Buffett song "Changes in Latitudes, Changes in Attitudes." With changes in elevation, come changes in precipitation. The elevation in North Texas generally slopes upward as you head west. For instance, Dallas sits at 430' above sea level. As you drive west, the elevation rises to 650' above sea level by the time you reach Fort Worth. Keep traveling west and the terrain becomes hillier as you continue to gradually gain elevation. Anyone who has traveled west on I-20 is familiar with Ranger Hill in Eastland County, which can be a treacherous spot to navigate in winter. The elevation at the base of the hill is roughly 1,100' above sea level but rises to just over 1,500' above sea level at the crest, approximately 1,000' higher than the average elevation in the DFW area.

Many of the high points in the western counties of North Texas sit at elevations above 1,500'. Recall that rising air cools. When air first begins to rise, it cools at the dry adiabatic rate of 5.5°F for every 1,000' of vertical movement. Once the air becomes saturated, it cools at the moist adiabatic rate of 3.3°F per 1,000' of vertical rise. So, if air is forced to rise another 1,000' as it is lifted over the higher terrain west of the Metroplex, it experiences additional cooling of roughly 3°F to 5°F. That can easily be the difference between 35°F and cold rain in Dallas and 30°F to 32°F and snow in Jacksboro and Bowie. The hills and valleys that become more prevalent as you head west and north of Dallas and Fort Worth can also aid in trapping the shallow cold air masses that often ooze southward into North Texas behind cold fronts. That colder air near the surface can favor freezing rain, while cold rain falls farther east and southeast.

THE 2009 CHRISTMAS EVE BLIZZARD

North Texas can go several years without seeing any significant snow or ice events. That winter weather hiatus ended in December 2009 with the first in a series of major storms that hit North Texas over the next four winters. The winter blitz began with a Christmas Eve storm that brought rare blizzard conditions and provided the entire DFW area with its first white Christmas since 1926. A powerful upper-level low, in tandem with an Arctic cold front, blasted areas from Jacksboro to Bowie to northwestern Cooke County with 9" of snow. The storm cranked up wind gusts over fifty miles an hour that generated whiteout conditions and snow drifts as high as 5' northwest of the Metroplex. The portion of US Highway 287 between Decatur and Wichita Falls was impassable from Christmas Eve into Christmas morning. Temperatures during the evening of December 23 were in the sixties in Dallas and Fort Worth, and less than twenty-four hours later a 3" blanket of snow covered the ground. Wind gusts of forty to fifty miles an hour during the late afternoon and evening blew the freshly fallen snow into blinding swirls that at times dropped visibility to near zero across the Metroplex on Christmas Eve. This was the first measurable snow on record for the DFW area and ensured that dreams for a white Christmas in North Texas would finally come true.

RECORD-BREAKING SNOWSTORM ON FEBRUARY 11–12, 2010

Just six weeks after the Christmas Eve blizzard, it appeared that conditions were aligning for another potential snowstorm, but none of the local meteorologists, including yours truly, foresaw what was coming. While one computer model showed the potential for more than 4", there was too much uncertainty regarding temperatures to forecast that with any degree of confidence. Snowfall events greater than 4" are a roughly one-in-ten-year event in North Texas. Going back to 1898, there have been twelve storms that have blanketed the DFW area with more than 4" of snow. By the time the snow had tapered off on February 12, 2010, 12.5" of snow had fallen at DFW Airport, establishing a new twenty-four-hour snowfall record.

How did meteorologists fail to forecast this record-breaking snowstorm? First, we anticipated the rain-snow mix to last much of the morning on February 11 before changing to all snow. At 4 a.m., the temperature at DFW Airport was 37°F with a mix of mostly rain and some snow. In less than two hours, the temperature dropped to near freezing and the precipitation switched to all snow several hours earlier than expected. We also failed to accurately predict the strength and speed of the upper air disturbance diving south from northwest into southwest Texas. This system was stronger

TOP: Deep Snow Blankets Cedar Hill *(Jeannine Shaw)* BOTTOM: 2010 Record Snowfall in Fort Worth *(Robert Howington)*

and tracking more slowly than most of the computer models had showed previously, and it had an impressive plume of subtropical moisture feeding into it. The disturbance also took on a "negative tilt." Instead of being aligned in the vertical, it leaned to the left in a ten to four o'clock orientation, forcing air to spread apart in the upper atmosphere, and inducing a more powerful lift and much higher snowfall rates. There were reports of "thundersnow," with lightning and thunder accompanying the last heavy bands that moved through during the wee morning hours on February 12. North Texans surely did a double take as they witnessed flashes in the sky and heard the rumbling of thunder in the dead of winter!

During the twenty-four to forty-eight hours leading up to the rare thundersnow, the computer models were highly inconsistent, leading to a low degree of confidence in their solutions. The February 10 evening run of the North American Model (NAM) came in cooler with more moisture, but surface temperatures were still borderline for more than 4" of snow, let alone 12". The Global Forecast System (GFS) model indicated more potential for cooling but showed less moisture. I went with a prediction of 2" to 4" the night before, with higher amounts possible north and west of DFW. In weather forecasting, you rarely see a meteorologist forecast an unprecedented, record-breaking event; it would be like betting all your money on the 100–1 longshot horse. There's a 1% chance you hit it but a 99% chance that you go home broke. This storm took the perfect track, with temperatures just cold enough to support snow through virtually the entire event and it had a lot more moisture to tap. The end result was a rare snowfall jackpot. Another late-season March snowstorm would make the 2009–2010 winter season the second snowiest on record with a whopping 17.1" of snowfall.

SUPER BOWL WEEK 2011

If there is ever a time you want to dial up Chamber of Commerce weather, it's the week you play host to the world's biggest sporting event. The week before the world descended upon the Dallas-Fort Worth area for Super Bowl XLV on February 6, 2011, the weather was superb. We had back-to-back 75°F days on January 28 and 29, followed by a pleasant 70°F day on January 30. Then an Arctic front plowed through North Texas on January 31 and delivered a blast of some of the coldest air we'd felt in decades.

AT&T Stadium, February 4, 2011 *(Doug Kovach)*

Temperatures were in the low fifties just ahead of the front at 1 p.m., and twelve hours later had plunged into the low twenties, with wind chills hovering around zero. North Texas remained locked in the deep freeze for the next four days, spending over a hundred consecutive hours below freezing from February 1 until the afternoon of February 5. To add insult to injury, there were two significant ice and snow events that cancelled hundreds of incoming flights, crippled transportation across the Metroplex, and prompted rolling blackouts to ease the burden on the power grid. Tom Bradshaw, Meteorologist in Charge for the National Weather Service Forecast Office in Fort Worth, admitted that, beyond presenting some huge forecasting challenges, the timing of the brutal winter barrage could not have been any worse. "From a marketing standpoint, it was one of the worst breaks Mother Nature could have dealt us."

The first round of wintry weather began as heavy rain ahead of the Arctic front on January 31, then transitioned to heavy sleet with thunder and lightning during the early morning hours of Tuesday, February 1. The "thundersleet" was blown sideways by winds that gusted to over fifty miles an hour at DFW Airport. As temperatures hovered around 20°F all day, the thick layer of sleet froze to rock-solid ice. Green Bay Packers and Pittsburgh Steelers fans, expecting to escape winter's chill for a few days in balmy Texas, instead felt like they had never left home. Unfortunately, this was only act one of Mother Nature's cruel two-act play. With Arctic air well entrenched, the stage was set for act two, which was delivered Thursday night into Friday.

Due to an apparent lack of moisture, the models showed little snow for North Texas. In fact, the early evening sounding (temperature and moisture profile at DFW Airport) showed that most of the atmosphere was dry. However, a strong upper-level low in El Paso was moving toward North Texas, and southerly winds ahead of it were quickly tapping into Gulf moisture. Snow started falling in the Waco area during the early evening hours of February 3 and then blossomed as it pushed north into the Metroplex late that night. The powdery snow fell steadily overnight into the morning hours of February 4. The storm was highly efficient in squeezing every bit of moisture out of the clouds, and with extremely cold road surfaces, all of the snow that fell accumulated. Snowfall totaled 5" at Dallas Love Field, 6" in parts of Collin County, and 7" to the northeast in Emory and Sulphur Springs.

A long, frigid week went from bad to worse on Friday, when a huge slab of snow cascaded off the roof of AT&T Stadium, injuring six contractors hired by the NFL to prepare the stadium for the game. The stadium avalanche left one of the workers in critical condition.

North Texas thawed out over Saturday and Sunday, with dry weather and temperatures that moderated into the fifties and made tailgating possible. The weekend ended with more precipitation. Fortunately, it was rain, not ice or snow. By the following Sunday, temperatures were once again in the seventies.

DECEMBER 5–6, 2013, "ICEMAGEDDON"

The ice storm dubbed "Icemageddon" was set up by a very large upper-level trough centered just north of Lake Superior. The deep trough extended over a thousand miles south into northern Mexico and placed North Texas under the influence of a belt of swift southwesterly winds aloft. This southwest flow brought in warmer air from 5,000' to 10,000' above the surface, producing what meteorologists refer to as a "warm nose." This particular warm nose was quite pronounced. At the surface, an early-season Arctic cold front was charging south toward North Texas. Ahead of the front, temperatures soared to 79°F on December 4 but plunged rapidly on December 5 as the front barreled through and kicked up robust north winds. Freezing rain began on the evening of December 5, with temperatures in the Metroplex in the low thirties. By the morning of December 6, we had tumbled into the mid- to upper twenties and the freezing rain had become steadier and heavier. With the Arctic pipeline flowing freely, the subfreezing

Collapsed Carport in Grapevine
(Diana Marie)

layer grew deeper, allowing the falling raindrops to become sleet. Heavy showers of sleet fell for several hours, and when it finally tapered off on the afternoon of December 6, as much as 5" of ice and sleet had accumulated in parts of North Texas.

This epic freezing rain and sleet event will always be remembered for introducing North Texans to the dreaded "cobblestone ice." A combination of tire compaction, partial melting into mounds of slush, sand treatment, and refreezing at night produced rocky formations of ice on roads throughout the Metroplex. My daily thirty-mile commute took me across the Highway 183 bridge that crossed the Elm Fork of the Trinity River. The cobblestone ice on that long stretch of roadway vibrated my entire car and rattled every bone in my body. A sixty-one-hour consecutive stretch of subfreezing temperatures kept portions of North Texas "iced in" for the next four days. Even after the mercury finally rose to freezing on the afternoon of December 8, crews took another day or two to clear the mammoth chunks of ice from area roads.

THE ICY BLUNDER ON THANKSGIVING DAY

November 25, 1993 was the coldest, snowiest Thanksgiving Day ever for the DFW area, with temperatures that dropped below freezing during the afternoon. The swirling, bitter cold winds were accompanied by snow and sleet that turned the annual Cowboys tilt at Texas Stadium into the football ice follies and made it one of the most memorable games in NFL history. Cowboys officials had borrowed new tarps from the Cotton Bowl to cover the field, but they became weighed down with so much ice that the grounds crew worried they might not be able to remove them in time for kickoff. With the aid of concession workers, the grounds crew scraped, shoveled, and swept as much ice as they could, and pried the tattered tarps from the turf. Meanwhile, showers of snow and sleet continued to fall through the opening in the roof of Texas Stadium, and in the hour before kickoff, the uncovered field looked like an ice-skating rink.

The Turkey Day showdown pitted the Cowboys against the Dolphins, and the game came down to the final fifteen seconds. Down by one, Miami went for a game-winning field goal, but Dallas blocked it. The stadium erupted and the Dallas sideline cheered to celebrate a Cowboys victory—or so they thought. That's when Cowboys defensive lineman Leon Lett, thinking the play was not over, stumbled after the ball, spinning like a top on the icy field. Moments later, NBC announcers Dick Enberg and Bob Trumpy would say the words that make Cowboys fans cringe to this day: "Someone touches the football. It's Leon Lett. No!" As Enberg and Trumpy made that now-famous call, more than sixty thousand Cowboys fans watched in disbelief at what was playing out on the "frozen tundra" at Texas Stadium. In his most regrettable play as a Cowboy, followed closely by his fumble in the previous year's Super Bowl, Lett accidentally kicked the football and Miami recovered it near the goal line with just three seconds on the clock. The Dolphins got a second chance at a field goal and converted, sending the Cowboys to a gut-wrenching 16–14 defeat.

THE 1978 NEW YEAR'S EVE ICE STORM AND THE CHICKEN SOUP GAME

The Thanksgiving Day Cowboys debacle wasn't the only holiday football game marred by untimely harsh weather. On New Year's Eve of 1978, North Texas was hit with its worst ice storm in three decades. With Arctic high pressure firmly entrenched across the central United States, moisture rode over the dome of frigid air at the surface. The freezing rain that resulted produced a hundred-mile-wide swath of ice that stretched from Waco to Paris. In the Metroplex, freezing rain fell for several hours, encasing the entire area in a sheet of ice up to 2" thick. The ice storm caused gridlock on the highways and left nearly three hundred thousand customers without power in Dallas County alone. Some homes were in the dark for up to ten days.

On New Year's Day, the 1979 Cotton Bowl game between the Notre Dame Fighting Irish and the University of Houston Cougars was played in bone-chilling conditions. Temperatures at kickoff were in the twenties, with wind chills that dipped below zero. A Zamboni was brought over from the minor league hockey facility next door to help clear the field, and fans in the stands used their shoes to break the ice off the aluminum seats. During the game, Notre Dame quarterback Joe Montana, who had been suffering from the flu, fought off hypothermia. With a body temperature that had dipped to 96°F, Montana was forced to retire to the locker room early. At halftime, the medical staff worked feverishly to warm Montana by covering him with blankets and feeding him chicken soup. Large vats of chicken bouillon had become a staple in the Irish locker room on cold days, and on this day, the old school remedy became the stuff of legends. With the Irish down 34–12, the legend himself, Joe Montana, returned with 1:41 left in the third quarter and led a furious comeback. On the last play of the game, Montana scrambled to avoid a sack, then hit his favorite target, Kris Haines, in the end zone for the game-tying touchdown. The Irish tacked on the extra point, escaping with a dramatic 35–34 victory.

CHAPTER 11

BE PREPARED!

April 10, 1979 is a day Donna Hanna will never forget. On that day, Wichita Falls took a direct hit from an F4 tornado that killed forty-five people and injured more than 1,800 others. The Wichita Falls tornado was the deadliest in a swarm of fifty-nine tornadoes that hit northwest Texas and western Oklahoma on "Terrible Tuesday." Donna was at her home in Altus, Oklahoma, watching live television coverage of the enormous tornado taking dead aim at Wichita Falls. "They were showing the tornado. There were three of them coming together and it was coming toward their TV station. Then suddenly they went off the air."

The multi-vortex tornado grew to become a mile-and-a-half wedge that consumed the entire horizon. The photo below was taken by Troy Glover, who worked at the maintenance department at Bethania Hospital in Wichita Falls, now the United Regional Healthcare System. Coincidentally, the hospital had just conducted its annual disaster drill earlier that day. "Part of the disaster plan for the hospital was for someone to go to the roof and watch for tornadoes threatening the hospital," says Glover. "So I volunteered to take the spotter role. That is how I happened to be on the roof with a camera when that tornado struck." Luckily, the tornado veered away from the hospital, but Glover watched in horror as it devastated entire neighborhoods.

TOP: Wichita Falls Tornado (Troy Glover) BOTTOM: Cars Flipped Like Toys *(NSSL/ Don Burgess)*

Over its eight-mile rampage through residential areas in Wichita Falls, the tornado flattened more than three thousand homes and one thousand apartments and condominiums, leaving at least twenty thousand people homeless. Of the forty-five deaths, twenty-five of them were vehicle related. Sixteen of those involved people who tried to escape the tornado in their cars, and eleven of those sixteen people left homes that did not get damaged.

Two weeks after Donna Hanna watched the tornado tragedy unfold live on the air, she traveled to Wichita Falls for an appointment at Sheppard Air Force Base. Donna says watching the tornado on television and seeing photos of the damage is one thing, but driving through the destruction was something entirely different. "I was stunned and awed. I grew up during the Vietnam War and it was like the city [had been] carpet-bombed. Houses were completely flattened. There was virtually nothing left standing." Seeing entire neighborhoods leveled and realizing how many lives would be forever changed by that tornado had a profound and lasting impact on Donna. She had always maintained an appreciation for the power of storms, but this would serve as a lasting reminder of the importance of heeding warnings and getting to a safe shelter as quickly as possible.

Donna Hanna's Damaged Home

Fast-forward to December 26, 2015. Donna was now residing in Rowlett, in a quiet neighborhood that she had called home for the past twenty-one years. Once again, she found herself glued to the TV set, but this time the tornado was much closer to home. "I had the TV on while I was working on a jigsaw puzzle. Then I received an alert on my iPhone to take cover immediately." Donna did exactly what she had programmed herself to do whenever there was a tornado warning. She grabbed her phone, her chair, and a book, and took shelter in her pantry underneath the stairs. She wasn't in the pantry for more than a few minutes when her oldest son called. "He said, 'Mom, they're saying that thing's right on top of you! Are you taking cover?'" She was able to respond to his message. "I told him I was in the pantry where I always am. About that time, it hit. Bang, bang, bang, bang, bang!" It lasted less than a minute. She thought it had only taken part of the roof, but when she emerged from the pantry, she was stunned by what she saw. "My front wall was [lying] in the yard. All the windows were blown out. The roof was gone. And right in the middle of the living room was a big beam. If I had not left my living room, that beam would have crushed me." Her home was left in ruins, but Donna escaped without a scratch.

STAY INFORMED

Mark Fox, the former warning coordination meteorologist for the National Weather Service Forecast Office in Fort Worth, surveyed some of the worst damage from the December 26 tornado outbreak and conducted dozens of interviews with storm survivors. The response he heard over and over again was, "We didn't know it was coming." Hearing those words was very troubling for him, especially considering that the threat of severe weather had been well advertised for days, and a watch and warnings were issued well in advance. The disconnect that left scores of residents in the dark about the impending dangerous weather underlines the importance of having multiple ways of receiving weather alerts. To keep yourself and your family safe, you can take the following precautions:

- Watch our live weather coverage on FOX 4 News, where our experienced team of meteorologists will track storms and alert you the minute dangerous weather threatens your vicinity.
- Keep up to date on our website at fox4news.com/weather.

- Follow us on social media via the FOX 4 Weather page and my Meteorologist Dan Henry page on Facebook. Likewise, you can get frequent updates @Fox4weather on Twitter.
- Download our free, versatile weather app, the WAPP, to get the latest forecast, watches, warnings, and live interactive radar.
- Have a NOAA Weather Radio at your bedside to receive overnight alerts.
- Make sure your phone is enabled to receive Wireless Emergency Alerts. The WEA messages are broadcast from area cell towers to your mobile device and will show the type and time of the alert.

I'm normally not one to advocate spending your hard-earned money on apps that clutter your phone, but I would highly suggest you check out RadarScope. This powerful app gives you access to 233 individual radar sites all across the United States, and as far away as Japan, and for a reasonable yearly fee will give you access to real-time lightning data. A NOAA Weather Radio can be a lifesaver too. These radios broadcast vital weather information from your local National Weather Service office 24/7 and have a battery backup so you can continue to receive alerts if you lose power. Many of these radios use Specific Area Message Encoding (SAME) technology to signal a loud tone and play audio of a computerized voice that reads the local weather alert. The bottom line is, the more ways you have to stay informed, the more time you will give yourself to take the necessary actions to keep yourself and your family safe.

KNOW WHERE TO GO AND GET THERE QUICKLY

Post-storm studies have shown that the vast majority of tornadoes and severe wind events are survivable if you promptly take shelter in an interior room on the ground floor of a well-built home. If you are one of the lucky few in Texas who have a basement or an underground shelter, that is the safest place to be in a tornado. For most of us, that safe shelter will likely be above ground and will vary according to where you live or where you are when the tornado hits.

At Home

The safest place to be is in an interior closet, bathroom, or pantry on the ground floor, away from exterior walls or windows. An interior bathroom can afford you additional protection because pipes help reinforce the walls and the bathtub can provide some added shelter. You should try to avoid being directly beneath the chimney or directly below heavy furniture, appliances, or a pool table on the upper floors.

Jim Joerger Watching FOX 4 Storm Coverage from his Safe Shelter

Apartment

If you're an apartment dweller, you will likely need to do some advance planning. If your apartment building has a basement or underground parking garage, go there. If you live on a higher floor, check with your landlord to see if there's a reinforced shelter on the property to use, or if the property's clubhouse or laundry room is an option. Alternatively, ask a neighbor on the first floor if you can ride out the storm there.

Mobile Home

Even well-anchored mobile homes will not offer you safe protection. Take shelter in a permanent, sturdy home, public building, or designated community shelter if there is one nearby. If there is not a more substantial structure you can get to quickly, seriously consider staying with a friend or a family member who lives in a permanent home. This may require you and your family to evacuate much earlier, preferably when a watch is issued. Waiting until a warning is issued to leave your mobile home will likely mean you are too late to get to safety and may put you at greater risk of being caught up in the storm.

Office

Go to an enclosed, windowless area in the center of the building and on the lowest floor possible. Interior stairwells are usually good places to take refuge and will allow you to get to a lower level quickly. Do not use elevators because the power may fail, leaving you trapped.

School

Go to an interior hall or windowless room in an orderly manner as directed. Crouch low, keep your head down, and protect the back of your head with your arms. Stay away from windows and large, open rooms like gymnasiums and auditoriums.

On the Road

Avoid being caught in this situation by delaying travel plans until the threat of severe weather clears. If a tornado is spotted in the distance while you are driving, do not seek shelter under a bridge or overpass. Head to the nearest substantial building, such as a truck stop, restaurant, or convenience store. In some situations, you may be able to drive away from the tornado, but this will depend on traffic, road options, weather conditions, and being able to clearly see the tornado and know what direction it is moving. If there is no escape possible, stay in the car with your seat belt on. Put your head down below the windows and cover yourself with a blanket if possible. If you can get significantly lower than the level of the roadway, leave your car and lie in that area, covering your head with your hands. Both of these are last-resort options that still leave you highly vulnerable.

EMERGENCY SUPPLY KIT

When a warning forces you to take shelter, time is of the essence. You may only have a few precious minutes to get everyone to safety. That is why keeping an evacuation kit in your safe shelter along with a few basic necessities will save you time and worry during a tornado threat. Many of the following items will fit in a backpack, so they can easily be transported. Your kit should include things like:

- **Food and Water:** Nonperishable foods that do not have to be cooked, a manual can opener, and at least a few gallons of water
- **Tools/Technology:** Flashlight with extra batteries, battery-operated radio, wrench or pliers (to turn off gas/water), air horn or whistle, and cell phone with charger
- **Medical:** First aid kit that includes bandages, cleansing wipes, antibiotic ointment, tweezers, and prescription medicines
- **Documents and More:** Credit card, cash, spare set of keys, birth certificates, insurance policies, ID cards, and household inventory. Keep these items in a waterproof container.
- **Clothing/Protection:** Hard-soled shoes, extra clothing, sleeping bags, blankets, and bicycle helmets (for head protection). If possible, store these items in your safe shelter.

Surviving the disaster is your first order of business, but having a well-stocked emergency supply kit will offer you peace of mind, and any one of these items may help you or save your life in the chaotic aftermath.

LESSONS LEARNED FROM JOPLIN

On May 22, 2011, one of the deadliest tornadoes in US history struck Joplin, Missouri, killing 158 people and injuring more than one thousand others. The EF-5 tornado was the first single tornado to result in more than a hundred fatalities since the Flint, Michigan tornado in 1953. A service assessment team, composed of experts from within and outside of the National Weather Service, examined the warning and forecast services provided before, during, and after the 2011 event. The team conducted nearly a hundred interviews with storm survivors, emergency managers, media, and first responders. The results of those interviews, included in a formal report released in July 2011, uncovered some sobering revelations. Despite excellent warnings that were issued thirty minutes prior to the tornado entering Joplin city limits, the majority of residents did not immediately take protective action. In fact, there were numerous accounts of people running to take shelter in their homes just as the tornado struck, despite significant advance warning of the risk. The comprehensive report cited the main reasons for this. "First of all, the response that struck us was the apparent apathy that respondents had concerning tornadoes prior to May 22. Generally, people felt that deadly tornadoes just didn't happen in that part of Missouri." That is simply an element of human nature that meteorologists struggle with. For many individuals, perception of risk is experience based. Mark Fox heard similar sentiments during his survey work of the December 26 outbreak in North Texas, such as, "Tornadoes don't happen in December, so I wasn't paying attention."

The Joplin tornado report also points to the frequent use of outdoor warning sirens creating a sense of complacency rather than a sense of urgency for those in the potential path of dangerous weather. On May 22, the National Weather Service issued the first tornado warning for northeastern Joplin at 5:09 p.m., and that was followed two minutes later by an initial three-minute activation of the outdoor warning sirens. A second tornado warning was issued at 5:17 p.m., which included the entire Joplin area. At 5:38 p.m., the outdoor warning sirens were activated again as the tornado approached the western city limits of Joplin. Despite the siren being sounded two separate times, the report cited confusion and an apparent cry-wolf syndrome as reasons why the outdoor warning sirens were not taken seriously. Some people thought the threat was over once the sirens ceased after the first alert. Many noted that they "hear sirens all the time" and the sirens "go off for dark clouds." Others told assessment team members that they are "bombarded with sirens so often that [they] don't pay attention" and that "all sirens mean is there is a little more water in the gutter."

CLEARING THE CONFUSION WITH OUTDOOR WARNING SIRENS

Outdoor warning sirens have also generated controversy and confusion over the years in North Texas. One of the more common misconceptions is that these are "tornado sirens." While the sirens are sounded for tornado warnings, they are also sounded for damaging winds, large hail, and other emergency situations. Also, as the name implies, the system is designed to alert people who are outdoors, and not to wake people up in the middle of the night. The criteria for activating the sirens are at the discretion of officials in each community and may change depending on special circumstances, such as big festivals that draw large crowds outdoors. As we've seen many times over the years, the decision of whether or not to activate the sirens generates controversy both ways.

On March 24, 2019, the outdoor warning sirens were sounded in Plano when a funnel cloud was spotted. The wail of the sirens was accompanied by a voice message of "tornado warning," even though the National Weather Service had not officially issued a warning. The message actually prompted many people to go outside to hear it—the exact opposite of what they were being urged to do. Because of the confusion, the city of Plano now uses "wail only" sirens. The city of Dallas was widely criticized for not activating its outdoor warning sirens in June 2019 during severe storms that packed eighty-mile-an-hour wind gusts that produced widespread damage to trees and power lines and led to a deadly crane collapse. The outdoor warning system is not perfect and serves as just one layer of warning coverage. When you hear the sirens, remain calm and get to safe shelter indoors. Then, tune to FOX 4, consult the relevant social media platforms, or listen to NOAA Weather Radio for more information and instructions.

HOW WIND EXPLOITS A HOME'S WEAKNESS

Even "weaker" tornadoes can cause extensive damage to your home. A high-end EF-1 tornado, packing winds up to 110 miles an hour, can collapse chimneys, break most windows, and remove portions of roofs. Ramp up the winds into the 111- to 135-mile-an-hour range, an EF-2 tornado, and well-constructed homes are shifted off their foundations while large portions of roofs are ripped away. Tim Marshall is a meteorologist and structural engineer who specializes in tornado-damage analysis. He says the wind will usually find the vulnerable point in a home: "Your home is like a chain, and the wind will exploit the weak link." When wind enters a home through a broken window or compromises a garage door, that wind acts on the inside of a home much like air blown into a balloon; it pushes on the walls and roof of the building from the inside. These forces within the structure, when combined with wind forces on the outside of the structure, often result in building failure. Tim says building codes don't account for all those stresses. "Building codes primarily take into consideration horizontal winds. That's a problem. Mother Nature has a very easy time taking roofs off of houses if you have a tornado with a lot of vertical velocity. It is exerting tremendous upward forces." Tim says the tornadoes that hit Dallas and Tarrant counties on April 3, 2012, were prime examples. "The EF-2 tornado that hit Lancaster was lifting tractor-trailers high into the air." Dozens of the five hundred trailers parked at Schneider National in Lancaster were tossed like toys, and some were vaulted hundreds of feet. Tim says while building codes have become more stringent in hurricane zones along the coast, building construction in inland areas has essentially remained the same in the forty-plus years he's been doing survey work. "There's a lot of politics involved in construction practices and improving building resistance. You have to keep housing prices low. The odds...of getting hit by a tornado [are very small], so they don't want to raise standards for everyone for odds that small."

INSTALLING A STORM SHELTER

When considering whether or not to install a safe room or storm shelter, you need to ask yourself a few questions: What are my existing refuge options? What level of safety am I comfortable with? What can I expect to pay for a quality storm shelter, and what can I afford to pay? The cost of a storm shelter varies with a number of factors, most notably the size of the shelter, the materials the shelter is constructed from, and whether you decide to locate it inside or outside your home. According to the National Storm Shelter Association, the cost for most shelters ranges from as little as $3,000 installed for a small, prefabricated safe room that measures ten square feet to $15,000 installed for a manufactured, residential storm shelter. FEMA (Federal Emergency Management Agency) has published a free booklet called *Taking Shelter from the Storm*, which includes several safe-room design options, siting considerations, and design requirements. The FEMA design guidelines for safe shelters are for 250-mile-an-hour winds, and the shelters must protect occupants

from wind-blown debris that an EF-5 tornado would generate. Research conducted at the National Wind Institute at Texas Tech University in Lubbock states that the shelter must be able to withstand impact by a two-by-four board (13.5' long and weighing fifteen pounds) "missile" traveling at one hundred miles an hour. Want to see what that looks like? Check out the FEMA video clip below.

Debris Impact Testing *(FEMA/Texas Tech University)*

Tim Marshall suggests ensuring that your shelter meets these standards and that it's anchored solidly to the foundation. "If it's an above-ground shelter, it needs to be bolted properly to a concrete slab. It should have a door that is solid steel with beefy hinges. It also has to have three locks to meet code." Federal funding is available for the construction of a storm shelter through the Small Business Administration, the US Department of Housing and Urban Development, and FEMA. To inquire about potential financial assistance for safe rooms, contact your state emergency manager or state hazard mitigation officer: https://www.fema.gov/safe-room-funding. In addition, funding may also be available through the North Central Texas Safe Room Rebate Program: https://saferoom.nctcog.org/.

For Jacqui Bloomquist-Taylor, the decision to build a storm shelter was all about peace of mind for herself and her family. She wanted a shelter large enough to accommodate herself and her husband; her mother, who lives across town; and her son's family, who live just one block away. In 2015, she invested $3,700, roughly the cost of a nice summer vacation, to build a shelter at her home on the shores of Lake Ray Hubbard in Rowlett. "My neighbors all thought I was crazy when I had an underground storm shelter installed in my backyard." She didn't imagine that less than six months later, on December 26, 2015, her underground bunker would be put to the ultimate test.

Just after 6:50 p.m. that evening, Jacqui's neighbor and tenant Jim Vincent warned the family members that a large tornado was heading their way. Jacqui and her entire family, as well as three neighbors and three pets, ran for cover in the backyard shelter. "We had twelve people, two dogs, and a cat with us. It was terrifying and it seemed like it lasted for an hour, but it was really less than five minutes. The lid of the shelter was clanking loudly, and you could hear the winds and debris hitting the shelter." When everything went silent, they slowly emerged from the shelter to discover their neighborhood in shambles. While Jacqui's home was still livable, her son's home was a wreck. "My son's house one street over was just a bunch of twisted framing, the roof was gone, and all the windows were blown out." Jacqui believes that had her son and his family remained at their home, they would have been injured or even worse. Looking back on her decision to invest in a storm shelter, she says, "It was worth every penny."

Rowlett Tornado Shelter Still Standing
(Tim Marshall)

Based on what he's witnessed over the past four decades while conducting more than ten thousand damage surveys from tornadoes, hurricanes, and hailstorms, Tim Marshall is also a big believer in storm shelters. He examined much of the damage in the Rowlett neighborhood where Jacqui lives, including one home just up the street, where nine people huddled together in the aboveground reinforced concrete shelter shown in the photo below. "The home was gone. The aboveground shelter was the only thing left standing. I'm certain those people would have sustained serious injuries or perhaps even been killed had they not gone into that shelter." There are thousands of homes with shelters in Central Oklahoma, where several devastating tornado outbreaks have occurred since 1999, and Tim sees an immense benefit. "There are hundreds of people alive and walking around today in Oklahoma City and Moore because they were in those shelters when the tornadoes hit. That's the untold story."

STORM SHELTERS NOW REQUIRED IN NEW-SCHOOL CONSTRUCTION

On May 20, 2013, an EF-4 tornado plowed into Plaza Towers Elementary School in Moore, Oklahoma. The school was leveled, leaving many students and teachers trapped beneath massive piles of debris. First responders pulled a car off of one teacher to discover three small children underneath her. The teacher and those three students survived, but seven of their classmates perished. The tragedy at Plaza Towers prompted city officials, school administrators, and parents to take action to make their schools safer. In 2015, a $209 million bond was approved to install storm shelters in every Moore school. That same year, changes in the International Building Code (IBC 2015) were adopted to require tornado shelters in all schools and first-response centers (new buildings and additions) with a greater-than-fifty-person occupancy in Oklahoma, Texas, and twenty-one other wind-prone states. Including these storm shelters in all new construction will be an expensive endeavor. FEMA estimates their complex design and building costs may run 30% higher than those of regular construction, but building shelters designed to withstand the forces of an EF-5 tornado will likely provide children greater safety at school than at home.

Several North Texas school districts have already completed or proposed projects that feature the new IBC storm shelters. Harmony Science Academy in Carrollton unveiled a new gym in 2018 that doubles as a storm shelter. Nearly 3.5 million pounds of concrete were poured to build the fortress, which features walls 15.5" thick, specially designed steel doors, and louvers that release pressure and are intended to prevent the roof from being lifted. Mesquite High School also included a new storm shelter as part of its $18 million expansion and renovation completed in 2018, and Carroll ISD in Southlake has built storm shelters at six of its eleven schools, with plans in the works to add them at its remaining campuses.

NATIONAL WEATHER SERVICE STORMREADY COMMUNITIES

StormReady, a nationwide program that helps communities better protect their citizens during severe weather, encourages communities to take a proactive approach to improving local hazardous-weather operations. The StormReady recognition is only given to communities with an established twenty-four-hour warning point and emergency operations center, as well as more than one method of receiving and distributing severe-weather warnings. Additionally, recognized communities must promote public readiness, have systems that monitor local weather conditions, and have emergency plans in place to guide residents to safety during hazardous weather incidents. Nearly a hundred Texas cities have received the StormReady recognition. Three of those cities—Cedar Hill, Desoto, and Duncanville— are in Tonya Hunter's jurisdiction. She is the regional emergency management operations coordinator for those southern Dallas communities, as well as for the city of Lancaster. Hunter says one of the important lessons learned from the 1994 Lancaster tornado was that emergency-management teams needed more eyes on approaching storms. "Cedar Hill sits up on an escarpment with good views of storms coming from the south or the west. We have ten Cedar Hill city employees who are trained as SkyWarn spotters, and we have specific locations where they go to observe and report what they are seeing." Their observations are sent out on a regional emergency operations channel (EOC) that is monitored by all four cities, as well as the 911 dispatch center.

These city-employee storm spotters work closely with volunteer storm spotters from the Southwest Dallas Amateur Radio Club, a community-service-oriented group that also monitors weather conditions throughout southwest Dallas County and northern Ellis County. Hunter says there is an emphasis on training city staff, their school districts, and the chambers of commerce so they are all on the same page and running like a well-oiled machine in those moments when the weather becomes severe. "Usually in spring, we'll do a severe weather or tornado exercise, having a scenario where a neighborhood or a school has been impacted. We're walking through a discussion-based exercise on responsibilities, what resources we have available, impacts to the community, and how we will respond [to] and recover from those types of incidents." More cities in North Texas continue to earn the StormReady designation every year, and the National Weather Service firmly believes that this local government/community partnership plays an integral role in the warning process.

CHAPTER 12

WHAT'S NEXT?

CLIMATE CHANGE AND
NORTH TEXAS WEATHER

C limate is what you expect; weather is what you get. Simply put, the difference between weather and climate is a matter of time. Weather is a day-to-day state of the atmosphere and its variation over short periods of time, ranging from minutes to weeks. Climate is the long-term average of the weather over a period of years or decades. As an example, the daily almanac we show on the air during our weathercasts displays the high and low temperatures for the day as well as the climatological averages. Those are the numbers I quote to brides and grooms, and their parents, when they ask for a six-month forecast in April for their upcoming wedding in October. And yes, we get asked all the time. I can give them an idea of the high and low temperatures for that day and those climatological averages. These long-term averages are also often referred to as "normals," which is ironic, because our everyday weather in Texas is anything but normal. The daily average high and low temperatures are based on thirty years of data. Why thirty years? In statistics, a sample size of at least thirty is generally considered necessary in order to get reliable estimates of the mean or average.

These thirty-year climatological averages are updated every ten years. The latest installment, the 1981–2010 period of record, added the warmest recorded decade in history, the 2000s. As a result, the average temperatures, or normals, have shown at least a small increase for many regions of the country. At DFW Airport, the 1971–2000 period of record showed an average annual temperature of 75.8°F, compared to 76.6°F for the 1981–2010 period of record. The average temperature in January increased 2°F and the annual precipitation increased from 34.73" to 36.14". The decade from 2011 to 2020 was even warmer than 2000 to 2010, which will bump up the normals even more. Texas State Climatologist John Nielsen-Gammon reports that, after a brief cooling trend in the 1970s, warming in Texas has outpaced the observed rise in global temperature. "Most parts of Texas are now about 1°F to 2°F warmer than the twentieth-century average temperature. Warming since the seventies for North Texas is a little more than the globe as a whole but is expected because land surfaces warm more than the oceans do." Are these increases concrete evidence of global warming or climate change? Not necessarily. To determine that, you need to take a much deeper dive into a substantially larger database that spans a much longer period of time.

GLOBAL WARMING VERSUS CLIMATE CHANGE

Let's begin by making a distinction between global warming and climate change. "Global warming" refers to the long-term warming of the planet since the Industrial Revolution. According to NASA's Goddard Institute for Space Studies, the average global temperature on Earth has increased by 1.9°F since 1880. The global increase in temperature from 1880 to present day has been recorded by many of the world's most established and well-respected meteorological organizations, including NASA's Earth Observatory, NOAA's Climate Prediction Center, the Japan Meteorological

Agency, and the Met Office Hadley Centre (United Kingdom). Two-thirds of the warming has occurred since 1975. This increase has been brought on by an excess of greenhouse gas emissions (carbon dioxide, nitrous oxide, methane, ozone, and water vapor) in the earth's atmosphere, mainly due to the burning of fossil fuels. "Climate change" includes the rising temperatures described by global warming but also encompasses all of the resulting effects. These include rising sea levels, the shrinking of the Greenland and Antarctic ice sheets, glacial retreat, the impact on ecosystems, and an increase in extreme-weather events.

HOW IS CLIMATE CHANGING?

The Intergovernmental Panel on Climate Change (IPCC) is a group of thirteen hundred independent scientific experts from all over the world under the auspices of the United Nations. The IPCC's *Fifth Assessment Report of the State of Climate Change* was issued in 2014. Its findings concluded, "Warming of the climate system is unequivocal, and since the 1950s, many of the observed changes are unprecedented over decades to millennia. The atmosphere and ocean have warmed, the amounts of snow and ice have diminished, and sea level has risen." The American Meteorological Society (AMS) is the nation's premier society of atmospheric scientists, with more than thirteen thousand members. The AMS issued its latest statement on April 15, 2019, in which it summarized the current best understanding of climate change. The findings indicate that global surface temperature has increased at a rate of 1.4°F per century from 1901 to 2017. That rate has accelerated significantly over the past decade and includes the five warmest years on record (2014–2018). This would result in a 3.4°F rise if it continued over the next century.

Climate change has also profoundly impacted the oceans. The sea surface temperature has warmed 1.8°F over vast portions of the oceans, sea levels have risen an average of 6.7" during the twentieth century, and the oceans have become 25% more acidic due to absorption of excess carbon dioxide.

Warmer global temperatures have led to a sharp decline in Arctic sea ice, which has experienced a 13% loss per decade since 1979. The overall effect of global warming on weather is more difficult to ascertain, but there is evidence that ocean warming is providing more energy to make hurricanes more intense, while the US average precipitation has increased by about 4% since 1900.

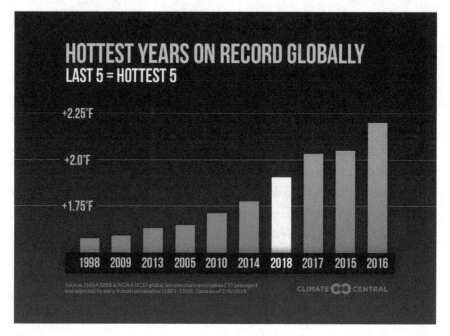

Hottest Years on Record *(Climate Central)*

SHRINKING ICE SHEETS AND RETREATING GLACIERS

On February 9, 2020, the temperature soared to 69°F at Argentina's Marambio research base on Seymour Island, just off the coast of the Antarctic Peninsula. It may be the warmest temperature ever recorded on the planet's coldest continent. The Antarctic peninsula is one of the fastest-warming regions on Earth, experiencing an average temperature rise of 5.4°F over the past 50 years. During the same 50-year period, the annual rate of ice loss has increased sixfold.

The shrinking of the Antarctic and Greenland ice sheets is an observable consequence of climate change. Together, the Antarctic and Greenland ice sheets contain more than 99% of the freshwater ice on Earth. Antarctic snowfall has accumulated over millions of years to build the world's largest ice sheet, essentially a massive glacier flowing in all directions. The Antarctic ice sheet extends almost 5.4 million square miles, roughly the area of the contiguous United States and Mexico combined. Scientists estimate global sea level would rise by about 200' if the Antarctic ice sheet were to melt. The Greenland ice sheet extends 656,000 square miles, covering most of the island of Greenland, and is roughly three times the size of Texas. Data from NASA's Gravity Recovery and Climate Experiment show a combined loss of over four hundred billion tons of ice per year between 1993 and 2016.

In contrast to the ice sheets, like those described above, which form over land, sea ice is simply frozen ocean water. It forms, grows, and melts back into the ocean it originated from, so one might initially think that its liquification would not have detrimental consequences. A striking visualization produced by NASA shows the seasonal variability of Arctic sea ice between 1984 and 2019. The animation shows younger ice, or first-year ice, in a dark shade of blue, while ice that is over four years old, called perennial ice, is shown as white. Perennial sea ice is the portion of sea ice that survives the summer melt season and can grow up to four meters thick during the winter as northern temperatures plummet. As the visualization clearly shows, perennial ice in the Arctic region is rapidly declining.

NASA Scientific Visualization Studio

How does melting sea ice influence our global climate? As warming temperatures melt sea ice, fewer bright surfaces are available to reflect sunlight back into the atmosphere. More solar energy is absorbed at the water's surface, ocean temperatures rise, and this becomes a vicious cycle of sorts. Warmer water temperatures delay ice growth in the fall and winter, and the ice melts faster the following spring, exposing dark ocean waters to sunshine and further warming for a longer period the following summer. It's not just sea ice that is vanishing; glaciers are retreating almost everywhere around the globe. The photos below show Muir Glacier in the Alaska Range in 1941 and 2004. Photographic evidence going back even further to 1892 shows that Muir Glacier retreated more than thirty-one miles between 1892 and 2004. Notice the prominent growth of vegetation on the mountain slopes in 2004, a stark contrast to the barren landscape of the 1941 image.

TOP: 1941: Muir Glacier, Alaska. National Snow and Ice Data Center, Boulder *(William O. Field)* BOTTOM: 2004: Muir Glacier, Alaska. US Geological Survey *(Bruce Molnia)*

RECONSTRUCTING PAST CLIMATE CONDITIONS

Modern instrumental temperature records date back roughly 150 years, so how do we know what temperatures were like on Earth in the centuries prior to that? Paleoclimatologists use what is known as proxy data to study and reconstruct past climate conditions. Proxy data are preserved physical characteristics of the environment such as tree rings, ice cores, fossil pollen, ocean sediments, and corals, all of which can extend our understanding of what the climate was like in the centuries before instrument measurements. Ice cores provide valuable information for reconstructing past climate. Scientists obtain these frozen time capsules by traveling to ice sheets, such as those in Antarctica or Greenland, and using a special drill that bores into the ice to extract ice cores. Particulates and dissolved chemicals that were captured by the falling snow become a part of the ice, as do bubbles of trapped air. With each passing year, distinct layers of ice accumulate, creating a unique record of the climate conditions at the time of formation. Ice cores allow scientists to glean a plethora of information about snow accumulation, local temperature, and the chemical composition of the atmosphere, including greenhouse gas concentrations. Volcanic and solar activity can also be determined. In 2017, scientists retrieved an ice core from the Allan Hills of Antarctica that contained ice 2.7 million years old.

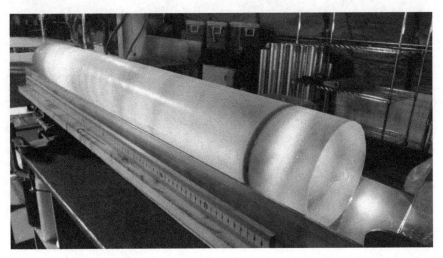

West Antarctic Sheet Ice Core *(Heidi Roop/NSF)*

WHAT'S NEXT? • 221

Bubbles in the ice were shown to contain greenhouse gases from Earth's atmosphere at a time when the planet's cycles of glacial advance and retreat were just beginning. The photo below shows an ice core from the West Antarctic Ice Sheet Divide. The dark band in this section of ice core is a layer of volcanic ash that settled on the ice sheet approximately twenty-one thousand years ago.

The graph below traces global temperatures over the past seventeen hundred years. It is shocking to see where we are today, especially when compared to the 1961–1990 global average. The temperature history is based on proxy data up until roughly 1850, and on modern thermometer-based data from 1850 through 2000. Over much of the past two millennia, the climate has warmed and cooled through the Medieval Warm Period and the Little Ice Age, but no previous warming episodes have been as large or abrupt as that of the past 120 years.

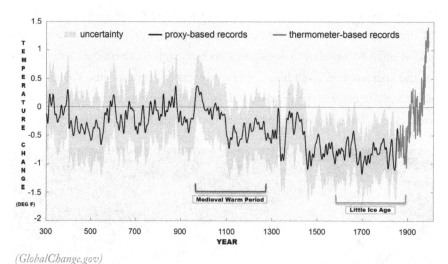

(GlobalChange.gov)

EARTH'S ENERGY BUDGET AND THE NATURAL GREENHOUSE EFFECT

One way to view the earth's energy budget is to compare it to your own household allocations. The difference between the money coming in and the money going out to pay bills determines whether you have funds left over or are left with a negative balance. In the earth-atmosphere system, the difference between energy coming in from the sun and that which is lost to space determines whether there is a net gain or loss of energy. If more energy is coming in than escaping, the earth's surface temperature will heat up; if more energy is leaving than coming in, the surface temperature will cool. Much of the incoming solar energy is composed of shorter wavelengths of visible light. Solar radiation entering our atmosphere is either reflected, absorbed, or transmitted (allowed to pass through). Of the sun's energy, 30% is reflected back to space by clouds, air molecules, and the earth's surface, while 19% is absorbed by clouds and atmospheric gases. The remaining 51% of the sun's energy passes through the atmosphere and is absorbed by the earth's surface. "Albedo" is the term used to define the portion of radiation that is reflected by a surface. Fresh snow reflects more than 90% of the sun's light, so it has a high albedo, while the oceans absorb much of the incoming light, giving them a low albedo.

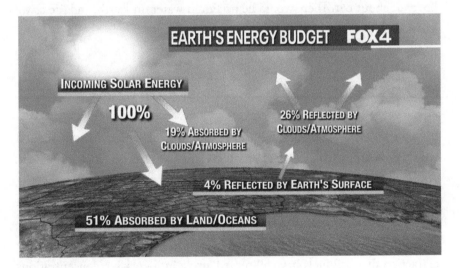

To balance the absorbed incoming energy, the earth must, on average, emit the same amount of radiation to space. The energy emitted by the earth is longer-wavelength infrared (IR) radiation, or heat energy. That loss of IR radiation happens not only during the day but long after sunset and through the night. Take, for example, the heating element of a stove. The radiation you feel being emitted from the heating element is mostly infrared, and it continues to emit heat energy long after the stove has been turned off. As mentioned, the earth emits IR radiation (heat energy) back to space. A substantial portion of that heat energy is absorbed by "greenhouse gases," including water vapor, carbon dioxide, methane, and nitrous oxide. These are called "selective absorbers" because they allow most of the sun's visible radiation to pass through but absorb much of the outgoing infrared radiation, preventing it from escaping to space. This causes warming to occur, commonly known as a "greenhouse effect." While the process is similar to what happens in a greenhouse, where solar radiation passes through the panes of glass, but IR radiation is trapped, there is an important distinction. Unlike warm air in the atmosphere, the warm air inside a greenhouse cannot mix at all with the cooler air outside. Nevertheless, even with mixing, greenhouse gases create a significant amount of warming. It's enough to warm the planet to a life-sustaining 59°F. In fact, if it were not for the greenhouse effect, the average temperature of Earth's surface would hover at about 0°F. There would be no liquid water on Earth, and life as we know it would not exist.

HUMAN EXPANSION OF THE GREENHOUSE EFFECT

Human activities, namely the burning of fossil fuels such as coal, oil, and natural gas, are enhancing the natural greenhouse that envelops our planet. All fossil fuels contain high percentages of carbon, so when they are burned, carbon combines with oxygen in the air to create carbon dioxide (CO_2). In addition to the combustion of fossil fuels, cement production, deforestation, the decomposition of waste in landfills, agricultural practices, and the extraction of fossil fuels have all played a role in increasing concentrations of greenhouse gases in the atmosphere. While water vapor is the most abundant greenhouse gas, it is usually an afterthought in the discussion of climate change because carbon dioxide and other long-lived gases are seen as the driving forces of climate change, while water vapor, which is continually being cycled through the atmosphere, is a feedback mechanism.

As air temperatures rise, more evaporation from the oceans and other water sources occurs, which leads to an increase in water vapor in the atmosphere. The higher water vapor content allows more thermal energy radiated from the earth to be absorbed, further warming the atmosphere. This is referred to as a "positive feedback" (an initial change causes a secondary change that in turn increases the effects of the initial change: a "snowballing effect"). Andrew Dessler, a climate scientist and professor of atmospheric sciences at Texas A&M, has conducted extensive research on climate change. He believes water vapor is an extremely significant feedback mechanism that more than doubles the direct warming from carbon dioxide. Adam Sobel, a professor of earth and environmental sciences at Columbia University, puts it this way: "CO_2 and other long-lived gases are the volume dial on the climate, and the water vapor amplifies the warming they produce." There is also a possibility that adding more water vapor to the atmosphere could produce a negative feedback effect by leading to more cloud formation, as clouds reflect sunlight and reduce the amount of energy that reaches the Earth's surface to warm it. In that case, adding more water vapor would have a cooling effect rather than a warming effect. The balance here is an active subject of climate science research.

Of the long-lived greenhouse gases, the three most impactful are carbon dioxide, methane, and nitrous oxide. Methane (CH_4) is a byproduct of the hydraulic fracturing (fracking) process for extracting oil and natural gas from underground and is also emitted by livestock farming and landfills. Nitrous oxide is a powerful greenhouse gas generated by the use of commercial and organic fertilizers and fossil-fuel combustion. While pound for pound, methane has more than thirty times the warming potential of carbon dioxide, carbon dioxide is over two hundred times more prevalent in the atmosphere. Carbon dioxide also lingers in the atmosphere much longer. While methane emissions take about a decade to leave the atmosphere, 40% of carbon dioxide will remain in the atmosphere for a hundred years after being emitted, while the final 10% will take ten thousand years to dissipate.

Carbon dioxide levels in the atmosphere are higher than they have ever been in recorded history. Scientists can confirm this with ice cores and paleoclimate evidence going back 800,000 years. During ice ages, CO_2 levels were around 200 ppm (parts per million), and during the warmer interglacial periods, they hovered around 280 ppm (see fluctuations in the graph). For millennia, CO_2 concentrations never exceeded 300 ppm. The Mauna Loa Observatory, on top of Hawaii's largest volcano, has been continually monitoring carbon dioxide concentrations for over sixty years. When measurements began in 1958, CO_2 levels averaged 315 ppm. Since that time, they have been steadily rising. In 2019, CO_2 levels set an all-time record, peaking at 415 ppm. The annual rate of increase since 1958 is about one hundred times faster than the upswing that occurred at the end of the last ice age roughly twelve thousand years ago. The unprecedented rise in carbon dioxide has mirrored the burning of fossil fuels since the Industrial Revolution and shows no signs of slowing down.

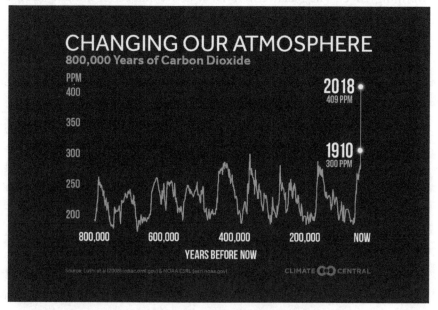

Carbon Dioxide Levels *(Climate Central)*

Understanding the extent to which climate change is man-made requires an understanding of Earth's powerful natural climate cycles, including changes in the sun's brightness, small variations in the shape of Earth's orbit and axis of rotation, large volcanic eruptions, and the El Niño–Southern Oscillation. Centuries of observations have shown that the number of sunspots waxes and wanes due to fluctuations in the sun's magnetic field in a cycle that lasts about eleven years. Sunspots are huge magnetic storms on the sun that show up as darker patches on the sun's surface. The small changes in energy emitted by the sun that occur during a solar cycle can exert a small influence on Earth's climate. Periods of intense sunspot activity are called solar maximums (2001) that can produce slightly higher temperatures, and solar minimums (1996 and 2006) can have the opposite effect.

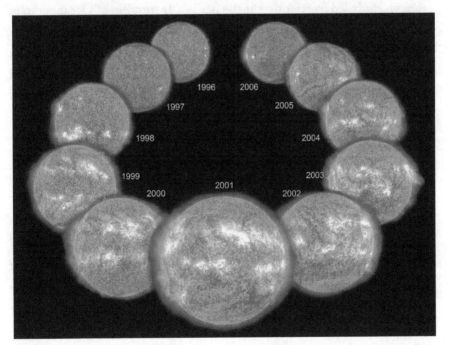

Images of the Sun During a Solar Cycle *(NASA)*

Sunspot activity also oscillates over several decades, a seesaw in which "grand solar minima" give way to "grand solar maxima." During the minima, which take place approximately once per century, solar output declines, sunspots vanish, and solar flares are rare. During the maxima, by contrast, the sun erupts with energy, and sunspots abound on its surface. Evidence suggests that the sun may be transitioning to another grand solar minimum. Could a decades-long period of unusually low sunspot activity produce enough of a cooling effect on Earth to curb global warming? The last grand minimum event, known as the "Maunder Minimum," occurred between 1645 and 1715 during a period called the "Little Ice Age." At this time, Earth's temperatures averaged about one degree cooler. Meteorologists from the UK, Finland, and the United States conducted a joint research study, published in 2017 and titled "The Maunder Minimum and the Little Ice Age: An update from recent reconstructions and climate simulations." They concluded that much of the cooling during the Little Ice Age was due to volcanic activity rather than a reduction in solar energy. Another extensive research study led by scientists at the Scripps Institution of Oceanography, at UC San Diego, examined the possible impacts of a "grand solar minimum." The simulations demonstrated that, while a dip in solar radiation could cool the earth's temperature, the results would only be temporary and the drop in temperatures wouldn't be significant, perhaps up to 0.5F, and would quickly rebound once the event concluded.

EARTH'S CHANGING ORBIT AND ITS EFFECT ON CLIMATE

During the last ice age, only twenty-one thousand years ago, thick sheets of ice stretched over Greenland and Canada and parts of the northern United States. The ice was over two miles thick over Hudson Bay and reached as far south as present-day New York and Cincinnati. For millennia, our planet has experienced cycles of glacier growth and retreat. These alternating glacial and interglacial periods coincide with variations in Earth's orbit called "Milankovitch cycles," named for the Serbian astrophysicist who calculated them in the 1920s. Changes in the shape of our orbit around the sun, as well as the tilt of Earth's rotation axis, and its wobble, take place over thousands of years. With these changes in Earth's movement and orientation, with respect to the sun, come changes in the amount of sunlight exposure.

The path of Earth's orbit around the sun is not a perfect circle, but an ellipse. Due to the gravitational pull of neighboring planets, particularly Jupiter, Venus, and Saturn, this shape changes from less elliptical (nearly a perfect circle) to more elliptical and back over the course of one hundred thousand years. When Earth's orbit is more elliptical, it travels farther away from the sun and receives less sunlight. At its closest point, in early January, called "perihelion," Earth swings to within 91,403,554 miles of the sun. In contrast, in early July, Earth reaches "aphelion," its most distant point, and is 94,513,221 miles from the sun. So, we are actually closer to the sun in winter than in summer. Thus, it's not our distance from the sun, but instead the tilt of our planet's axis that drives winter and summer.

We know Earth is spinning around its own axis, which is why we have night and day. However, this axis is not upright, and the angle at which Earth tilts varies according to a forty-one-thousand-year cycle and fluctuates between 22.1 and 24.5 degrees. Currently, the axis is positioned at about 23.5 degrees, which is roughly in the middle of the cycle, and is in a decreasing phase. At higher angles, Earth is tilted more toward the sun, and the seasons become more exaggerated as Earth receives even more sunlight during summer and less during winter. Aside from the tilt, the axis also wobbles like a top due to gravitational forces from the sun and moon. A complete wobble cycle is roughly twenty-six thousand years.

The collective effects of these changes in Earth's movements alter the amount of sunlight that reaches our atmosphere. At the Arctic Circle, sunlight exposure can vary by as much as 25%. When there is less sunlight during the summer months, temperatures are cooler, and more ice and snow remains. Over the course of many centuries, ice builds up and eventually produces an ice sheet. The average length of glacial periods has changed over time. The cycles range from roughly forty thousand years, which is closely aligned to changes in the tilt of Earth's axis, to roughly one hundred thousand years, coinciding with changes in the shape of our planet's orbit. So where are we today in the Milankovitch cycles? Earth's Northern Hemisphere currently experiences summer at aphelion (farthest from the sun), the planet's axis is in the decreasing phase of its cycle, and we are in the nearly circular part of our orbit. Given all of this, the current position of Earth's orbit should result in cooler temperatures, but instead average global temperatures are on the rise. The bottom line is that orbital phases can help explain very long-term changes in our climate, but they are not the culprit for our current warming trend.

VOLCANIC ERUPTIONS AND THEIR IMPACT ON CLIMATE

History has shown that major volcanic eruptions can cause climate changes for a period of months to up to a few years. In 1991, Mount Pinatubo erupted, sending a gigantic cloud of rock, ash, and gas more than twenty-two miles into the sky above the Philippines. In 1982, Mexico's presumed-dormant volcano El Chichon produced three eruptions in a little under a week and ejected twenty million tons of sulfur dioxide into the atmosphere. These particular eruptions, rich in sulfur gases, combine with water vapor in the presence of sunlight to create tiny, reflective sulfuric acid particles that grow in size and form a dense layer of haze. Major eruptions alter the earth's energy balance because volcanic aerosol clouds scatter a significant amount of the incoming solar radiation and absorb infrared radiation emitted by the earth.

As a result of the Pinatubo eruption, large parts of the planet cooled as much as 0.7°F the following year, in 1992. Researchers with the Max Planck Institute for Meteorology and Rutgers University ran global-climate-model simulations for the two years following the Pinatubo eruption. By comparing simulations with and without aerosols, the researchers found that the climate model calculated a general cooling of the global troposphere (lowest layer of the earth's atmosphere) consistent with what occurred during 1992 through 1993. However, the model simulations also displayed a clear winter warming pattern of surface air temperature over Northern Hemisphere continents. Man-made emissions can make the consequences of volcanic eruptions on the global climate system more severe. For instance, chlorofluorocarbons (CFCs) in the atmosphere start a chain of chemical reactions with volcanic aerosols that can deplete the ozone layer in the stratosphere. Stratospheric ozone provides a protective shield against the sun's harmful ultraviolet rays. Volcanic eruptions like Pinatubo can temporarily alter global climate and even change atmospheric circulation patterns, but the extent to which this occurs will continue to be an ongoing debate.

EL NIÑO AND LA NIÑA

Along the west coast of South America, the typically cool water supports large fish populations and anchovies that make up the largest fishery on earth. The cool water is rich in nutrients, providing bountiful harvests for the Peruvian fishing industry. The term "El Niño" ("The Little Boy," referring to the Christ child) was first used by Peruvian fishermen in the late nineteenth century to refer to the periodic emergence of unusually warm, nutrient-poor water off the coasts of Peru and Ecuador that chased the fish away. The name was chosen because the arrival of these warm waters usually occurred around Christmas. El Niño and La Niña ("The Girl Child") are opposite phases of ENSO, which is short for El Niño–Southern Oscillation. Operating in the tropical Pacific Ocean, ENSO is the earth's single-most influential natural climate pattern. While ENSO has occurred for tens of thousands of years, only since the second half of the twentieth century have meteorologists identified it and begun to understand its impacts.

The simplest way to understand El Niño and La Niña is to imagine warm water sloshing back and forth in the tropical Pacific Ocean. The top layer of the tropical Pacific (roughly two hundred meters in depth) is warm, with water temperatures in the seventies and eighties. Beneath this warm top layer, the ocean becomes much cooler and is far more static. The prevailing trade winds usually blow from east to west, which pushes the warm surface waters of the tropical Pacific Ocean toward Indonesia and Australia. This allows cool water to well up from the depths of the ocean along the coast of South America. During an El Niño, the easterly trade winds weaken and can even turn around into westerlies, allowing an immense swath of warm water to slosh from the western Pacific toward the Americas. The warm water

NASA Goddard
Space Flight Center

smothers the upwelling of cooler, nutrient-rich waters from the deep. Often, El Niño is followed by La Niña. During La Niña, the easterly trade winds strengthen, and warm water sloshes back in the opposite direction toward Indonesia and Australia. This 3-D visualization from NASA Goddard Space Flight Center chronicles the entire life cycle of the 2015–2016 El Niño event.

The circulation of the air above the tropical Pacific Ocean responds to this tremendous redistribution of ocean heat, causing precipitation patterns to drastically shift, with flooding rains in some regions and drought in others. Here in the United States, El Niño stokes an active storm track from Southern California through Texas to Florida, while dry, mild weather reigns across the north. La Niña episodes feature colder and stormier weather across the north, while mild and dry weather prevails across the south.

El Niño events occur roughly every two to seven years, as the warm cycle alternates irregularly with its cooler sibling, La Niña. Both phases typically peak during the winter months, slowly revving up months in advance. Though ENSO events are not caused by climate change, they undoubtedly enhance its impacts. The 1982–1983 El Niño is widely regarded as the strongest ever recorded. It incited widespread flooding in the southern United States and devastating droughts and forest fires in Australia and Indonesia. Another strong El Niño in 1997–1998 was a contributing factor in 1998 becoming the warmest year of the twentieth century.

METEOROLOGISTS AND CLIMATE SCIENTISTS
STILL DISAGREE BUT NOT AS MUCH

As a scientist, I thoroughly enjoy doing research, gathering facts, and crunching numbers. That is especially true when it comes to rendering an opinion on an important and controversial topic. Even among American Meteorological Society members there is still some debate, but the numbers of those who believe human activities are playing a major role in global warming are steadily rising. In a 2016 survey of AMS members, the majority of whom are meteorologists or atmospheric scientists, 67% said climate change over the last fifty years is mostly or entirely caused by human activity. Among the climate scientist community, 97% believe that humans are causing global warming and climate change. Why the disparity? Part of this may have to do with model skepticism. Meteorologists use computer models that are sensitive to small atmospheric changes; tiny discrepancies can compound over time, leading to big errors and little forecast accuracy beyond seven days. Consequently, they tend to be skeptical about long-term predictions made by global climate models. Meteorologists also live in the short term, observing and forecasting weather day-to-day. What will the high temperature be, and will it rain in Fort Worth tomorrow? Climate scientists look years and decades into the future and ask: How much warmer will the earth be fifty to a hundred years from now? How much will sea levels rise? In the search for answers, climate scientists analyze many other pieces of data, each a small piece of the complicated puzzle. How does the warming of our oceans spur a series of chain reactions that magnifies Arctic ice melt? How much methane is being released into the atmosphere by the sudden thawing of permafrost in Siberia and Alaska? Climate scientists believe they have assembled enough pieces of the puzzle to say that global warming and climate change is unequivocal.

CLIMATE CHANGE: WHAT DOES THE FUTURE HOLD?

In predicting the climate of the future, the biggest unknown is what actions humans will take. How will we change our use of fossil fuels and thus our

CO_2 emissions? There are other important factors to consider as well: world population and economic growth, energy consumption and sources, and land use. This is an extremely complex issue with hundreds of scenarios that can play out, but researchers have worked out four plausible scenarios that they believe represent the range of possible outcomes for future warming in the twenty-first century. These outcomes are referred to as Representative Concentration Pathways (RCPs). RCPs are scenarios that describe the possible trajectories for carbon dioxide emissions and the resulting atmospheric concentrations from 2000 to 2100. They also encompass the array of possible climate policy outcomes for the twenty-first century.

Each RCP provides us with a number that describes how the climate will fare in the year 2100. Generally, a higher RCP number describes a scarier fate. The RCP 2.6 scenario can be described as the "best case" for limiting human-induced climate change. It assumes a major turnaround in climate policies, that biofuel use is high, that renewable energy (solar and wind) use increases, and that global CO_2 emissions peak by 2020 and decline to around zero by 2080. It also assumes that the global population peaks mid-century at just over nine billion and that economic growth is high. Under this scenario, computer models project that the earth's average temperature will rise about 1°C to 2°C (1.8°F to 3.6°F) by 2100. At the other end of the spectrum is the RCP 8.5 scenario, often referred to as the worst case or "nightmare scenario." RCP 8.5 assumes that concentrations of carbon dioxide in the atmosphere accelerate and reach 950 ppm, more than double the current levels, by 2100 and continue increasing for another hundred years. It also assumes that global population growth is high, reaching twelve billion by the end of the century. Under this scenario, the average global temperature would likely rise 4°C to 5°C (7°F to 9°F) by 2100.

WARMER AND WETTER WORLD

Even an apparently small shift in global average temperatures results in a large change in extremes. Imagine all the days in a year. They typically range from very cold to very hot, but most days lie somewhere in the middle. On a graph, this makes a bell curve, with the average temperature located directly below the peak of the curve. In a warmer climate, the entire temperature distribution (second bell curve) shifts warmer, spelling more frequent heat and record-setting heat waves, and fewer outbreaks of extremely cold weather. Climate change is predicted to impact regions differently. For example, temperature increases are expected to be greater on land than over oceans, and greater in Arctic regions than in the tropics and mid-latitudes. As the world continues to warm, increased evaporation from soils will lead to more frequent and prolonged drought. Hotter days and warmer nights will increase heat stress on the body and raise the risk of heat-related illnesses. Likewise, air conditioning demands would escalate, stressing the electric power grid and driving up utility bills.

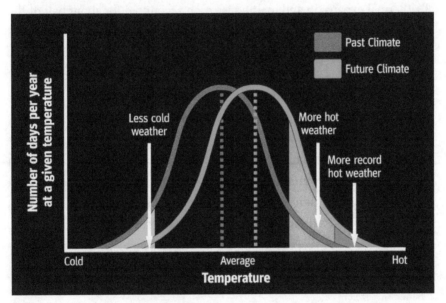

(Australian Climate Commission Report)

Another consequence of warmer global temperatures relates to precipitation. A warmer climate will cause higher rates of evaporation, accelerating the water cycle. More water vapor in the atmosphere will lead to increased precipitation and more extreme rainfall events. According to climate models, global average precipitation will most likely increase by an average of 3% to 5%, with a minimum increase of at least 1% and a maximum increase of about 8%, yet changes in precipitation will not be evenly distributed across the globe. Some regions will receive more snow, while others will see less rain. Some places will experience wetter winters while others will endure drier summers.

RISING SEA LEVELS

A warmer climate causes sea levels to rise via two processes. First, the melting of glaciers and massive ice sheets that cover Greenland and Antarctica add water to the oceans, thus raising sea levels. Second, water expands as it warms, so warmer oceans will increase in volume and raise sea levels. During the twentieth century, sea levels rose about twenty centimeters, or 8". Thermal expansion and melting ice each accounted for about half of the rise, though there is some uncertainty about the exact magnitude of contribution from each source.

Projecting future rates of sea-level rise is challenging due to the complexity of ice sheet dynamics. This involves the movement of large bodies of ice, as well as glacier calving, a phenomenon in which massive chunks of ice break off, destabilizing glaciers and exposing their interiors to warm

water and accelerated melting. Video footage (which can be viewed by scanning the QR code below) shows an extraordinary calving event in which a slab of ice the size of Lower Manhattan, and up to 3,000' thick, breaks off the Ilulissat Glacier in Greenland.

Mammoth Glacier
Breaks Apart
(Exposure Labs/
James Balog)

Even the most sophisticated climate models cannot simulate rapid changes in ice sheet dynamics and thus are likely to underestimate future sea-level rise. The general consensus of the world's leading ice sheet experts suggests a range of additional sea-level rise from about one meter (3.3') to as much as two meters (6.5') by 2100, depending on future greenhouse gas emissions and further warming. Presently, a majority of coastal communities experience five days of high-tide flooding annually, but by 2050 that will increase six-fold, amounting to approximately thirty days per year. In Galveston County, where much of the area is only a few feet above sea level, more than four thousand homes will be at risk for yearly coastal flooding by 2050. The higher water levels could spell doom for vulnerable cities like Miami, New Orleans, Boston, and the Houston/Galveston area when hurricanes and nor'easters come calling. Sea-level rise will enable these powerful storms to swamp areas much farther inland, submerging buildings, threatening lives, and dramatically accelerating shoreline erosion. Saltwater intrusion due to rising seas will also damage coastal habitats and estuaries and contaminate drinking water by infiltrating aquifers and reservoirs.

MORE INTENSE STORMS?

While citing global warming as the sole culprit for catastrophic hurricanes and tornado outbreaks makes for great headlines, this is simply not accurate. Global warming does not cause individual weather events, but it can potentially change their intensity and impact. The correlation between global warming and hurricane activity may be stronger than that between global warming and tornadoes. Although there may not be more tropical cyclones worldwide in the future, some scientists believe there will be a higher proportion of powerful and destructive storms, due to warmer seas. The higher ocean heat content provides more energy for hurricanes to tap and may enable those storms that do develop to be longer lasting. Currently, the area of the globe where tropical cyclones occur has expanded, and tropical cyclones are, on average, reaching their peak intensities in areas farther outside the tropics than a few decades ago. This trend may continue as sea temperatures rise farther away from the equator.

Scientists are more confident that the heavy rainfall and storm surge flooding associated with hurricanes will both increase. Several recent studies examined the extreme rainfall (40" to 60") produced by Hurricane Harvey in 2017 that led to catastrophic flooding. While Harvey's extraordinary rainfall was primarily due to the storm's slow movement over Southeast Texas, computer model simulations showed that the intense downpours were enhanced by 15% due to global warming, and that extreme rainfall events along the Gulf Coast are on the rise.

Finally, we know that rising sea levels are a product of global warming, which means coastal flooding problems can be worse when tropical cyclones make landfall. The link between global warming and severe weather is much less defined. While global temperatures have risen nearly 1°F since 1950, the number of strong and violent tornadoes each year in the United States hasn't changed much since then. The number of weak tornadoes (EF-0, EF-1) has increased dramatically in that same time period, but that is likely due in large part to better detection, which is the result of a dense network of Doppler radars, more people living in areas that were previously undeveloped, and readily available technology for documenting occurrences—technology as "basic" as our iPhones and Androids.

IMPACTS ON LIFE, AGRICULTURE, AND ECOSYSTEMS

Climate change is expected to have far-reaching impacts on human health, wildlife, agriculture, and water supplies. The changing environment is expected to cause an increase in waterborne diseases, poor air quality, a decline in water supplies, and the spread of illnesses like West Nile virus and Lyme disease transmitted by mosquitoes and ticks. Human health is vulnerable to climate change. Extreme heat events will result in increased hospital admissions for heat-related illnesses as well as cardiovascular and respiratory disorders. Scientists estimate that each increase of approximately 1.8°F (1°C) in summer temperature increases the death rate for elderly people with chronic conditions by 2.8% to 4%.

Ecosystems are also impacted by climate change. Temperature changes will alter the natural ranges of many types of plants and animals, shift the timing of events such as flowering and egg laying, spur changes in patterns of seasonal breeding, cause significant disruptions to the food web, and threaten some species with extinction. Increased heat, drought, and insect outbreaks are linked to climate change and will lead to reduced agricultural yields. Changes will be observed in the lengths of growing seasons, geographical ranges of plants, and frost dates. Finally, oceans will become saturated with CO_2 and will be unable to serve as a buffer by absorbing CO_2 that would otherwise stay in the atmosphere. At this point, human-induced emissions of CO_2 will all remain in the atmosphere, increasing the rate of greenhouse warming.

WHAT DOES CLIMATE CHANGE MEAN FOR TEXAS?

It is hard to say precisely how climate change will affect weather in Texas over the remainder of the twenty-first century, but temperatures will undoubtedly be warmer than in the past, and we will experience even more dramatic swings from drought to deluge. According to the Fourth National Climate Assessment Report released by the US Global Change Research Program in November 2018, annual average temperatures in the Southern Great Plains are projected to increase by 3.6°F to 5.1°F by the mid-twenty-first century, and by 4.4°F to 8.4°F by the late twenty-first century, compared to the average for 1976–2005. The lower end of those ranges reflects a more moderate RCP 4.5 scenario, while the upper end depicts a worst-case RCP 8.5 scenario. Future levels of greenhouse gas emissions will ultimately determine how much and how quickly temperatures will rise. To put that into perspective, let's assume the average January high temperature at DFW Airport jumps 4.5°F by 2050 and 6.5°F by 2100, roughly in the middle of the projected increase. That would mean the average high in January at DFW Airport would vault from 56.5°F to 61°F, which is just shy of Austin's current average January high of 61.5°F. By 2100, DFW Airport's average January high would rise another two degrees to 63.0°F, the same as Houston's current average high in January. The report also warns that extreme heat will become more common: "Temperatures similar to the summer of 2011 will become increasingly likely to recur, particularly under higher (greenhouse gas emissions) scenarios. By late in the 21st century, if no reductions in emissions take place, the region is projected to experience an additional 30 to 60 days per year above 100°F than it does now." The impact of climate change on the frequency and intensity of severe weather, such as tornadoes and hailstorms, remains difficult to quantify. Texas State Climatologist John Nielsen-Gammon says, "There's very limited evidence as to how warmer temperatures will affect severe weather partly because we have lousy historical information on tornadoes, so it's hard to compare the past to the present. Plus, you can't simulate those small-scale storms that produce severe weather using global climate models. Research tends to point to a possible increase, but I wouldn't put much faith in that."

Since warmer air can take up more water vapor before it becomes saturated, the atmosphere will be more conducive for producing rainfall. While climate change will likely result in average annual precipitation totals rising slightly, the overriding theme will be one of more "feast or famine" extremes. The Climate Assessment Report states, "the expected increase of precipitation intensity implies fewer soaking rains and more time to dry out between events, with an attendant increase in soil moisture stress. Studies that have attempted to simulate the consequences of future precipitation patterns consistently project less future soil moisture, with future conditions possibly drier than anything experienced by the region during at least the past one thousand years." More intense heat and lower soil moisture will lead to more prolonged droughts and increase the risk of wildfires. On the flip side, when the skies do open up, moisture-laden clouds will be capable of producing more intense rainfall rates and the likelihood of more frequent flash flooding events.

CLIMATE CHANGE: WHAT'S THE VERDICT?

For many years, I sat on the fence when asked to offer my opinion on climate change. I have listened to the arguments on both sides and carefully examined the research and the data. Since 1978, satellites have allowed scientists to make precise measurements of the amount of energy arriving from the sun. Total Solar Irradiance (TSI) is a measure of this solar power over all wavelengths. While TSI varies by approximately 0.1% due to sunspot activity, it has not increased since 1978, and therefore cannot explain the rapid warming of the past four decades. Scientists can also differentiate between CO_2 molecules that are emitted naturally by plants and animals and those that are a product of fossil-fuel combustion. Deforestation and the burning of fossil fuels produce lighter carbon isotopes than those from other sources. Scientists measuring carbon in the atmosphere can see that lighter carbon molecules are increasing, corresponding to the rise in fossil-fuel emissions. Evidence from proxy data, including ice cores, tree rings, sedimentary rocks, ocean sediments, and coral reefs, shows that the current warming is occurring ten times faster than it did when the earth emerged from the ice ages, and at a rate not seen in the last thirteen hundred years.

Finally, while climate models are far from perfect, they are one of the best tools for understanding the causes of observed climate change and forecasting future warming. The simulations run by climate models are very telling. When input data only from natural climate variations, such as the sun's intensity, changes in the earth's orbit, and El Niño, are used, the climate models cannot reproduce the warming in the oceans and the atmosphere that is occurring. Only when the emissions from human activity are included with the rest of the input data do the models accurately depict the warming we are witnessing. When you consider all of the evidence, it is hard to deny that climate change is real, and that mankind shoulders a large portion of the blame for the warming we have observed since the Industrial Revolution. What still remains up for debate is how much warming we will see in the future and how rapidly those changes will advance.

FINAL THOUGHTS ON CLIMATE CHANGE

Nearly all climate scientists agree that slashing carbon pollution will not, by itself, be enough to prevent the earth from overheating. That is why a secondary approach to tame global warming is gaining traction but also stirring up controversy. "Geoengineering" proposes using emerging technologies to siphon greenhouse gases from the atmosphere and shade the earth from solar radiation to partially offset the impacts of climate change. There are a number of geoengineering schemes, but many are seen as cost prohibitive and may produce unintended consequences. A few of the more intriguing proposals include chemically capturing carbon dioxide from the air ("carbon geoengineering"), spurring growth of carbon-eating plankton in the oceans, and injecting aerosols into the stratosphere to reflect sunlight, a concept referred to as "solar geoengineering."

Of these, solar geoengineering may demonstrate the most promise: Harvard University has a research program devoted entirely to this science and technology. This approach involves spewing sulfate aerosols or perhaps calcium carbonate into the upper atmosphere, where the tiny particles could reflect incoming sunlight back into space. Volcanic eruptions that spew ash miles into the atmosphere have had significant cooling effects on the planet, which offers some scientists hope that this strategy would work. Other scientists argue that the cooling effects of the artificially generated, sunlight-dimming haze could come at the cost of disrupting weather patterns and eating away at the ozone layer.

The most obvious way of shrinking our global carbon footprint is to cut back on our use of fossil fuels such as coal and oil. Harnessing the power of the sun and tapping into the energy of the wind can also help reduce greenhouse gas emissions. Technological advances have made solar panels smaller, cheaper to manufacture, and more efficient, leading to surges in solar installations across the world. When it comes to wind energy, Texas is at the forefront and has become the biggest generator of wind power in the United States. The Lone Star State produces nearly 20% of its electricity from wind, and that percentage continues to grow every year. Other ways of curbing our carbon binge include making improvements to energy efficiency and vehicle fuel economy, using biofuels from organic waste, building better batteries to store renewable energy, and protecting our forests.

Nearly eight billion people live on the planet, and if we intend to combat climate change, each of us will need to reduce our carbon footprint by consuming less and wasting less. Each individual's contribution may seem like a drop in the bucket, but our combined efforts can make a big difference.

		JAN	FEB	MAR	APR	MAY	JUN	JUL	AUG	SEP	OCT	NOV	DEC	YEAR
Temp (°F)	High	56.4	60.4	68.2	76.3	83.6	91.3	95.6	**96.1**	88.4	78.3	66.9	57.1	76.6
Record (Year)	High	93 (1911)	96 (1904)	100 (1916)	101 (2006)	107 (1927)	**113 (1980)**	110 (1998)	112 (1936)	111 (2000)	106 (1951)	94 (2017)	90 (1951)	113 (1980)
Temp (°F)	Low	**35.5**	39.4	47.0	54.8	64.2	71.4	75.1	75.1	67.5	56.8	46.2	37.1	55.8
Record (Year)	Low	-2 (1949)	**-8 (1899)**	10 (1943)	29 (1989)	34 (1903)	48 (1903)	56 (1924)	55 (1915)	40 (1942)	24 (1917)	19 (1950)	-1 (1989)	-8 (1899)
100°F	High	0	0	0	0	0.2	1.2	6	**9.2**	1	0	0	0	17.6
32°F	Low	**12.1**	6.8	2	0.1	0	0	0	0	0	0.1	2.3	9.7	33.0
RH	Avg	68%	67%	65%	64%	**70%**	67%	61%	59%	64%	66%	69%	**70%**	66%
	%	52%	54%	58%	61%	57%	67%	**75%**	73%	67%	62%	57%	52%	61%
Fog	Days	**2.1**	1.2	0.9	0.5	0.3	0.1	0.1	0.1	0.1	0.7	1.3	**2.1**	9.5
Storm	Days	1.4	2.0	4.0	5.7	**7.6**	6.3	4.7	4.6	3.3	3.2	1.9	1.3	46

DALLAS–FORT WORTH CLIMATE DATA: WIND, RAIN, SNOW

		JAN	FEB	MAR	APR	MAY	JUN	JUL	AUG	SEP	OCT	NOV	DEC	YEAR
Wind (mph)	Avg	10.8	11.3	12.1	**12.2**	11.4	10.5	9.9	8.7	8.6	9.7	10.6	10.3	10.5
	Dir	190°	170°	**170°**	170°	170°	170°	170°	170°	170°	170°	190°	190°	190°
	Max Gust	55	78	**79**	75	67	64	64	60	52	54	54	60	79
Rain (in)	Avg	2.13	2.66	3.49	3.07	**4.90**	3.79	2.16	1.91	2.55	4.22	2.71	2.55	36.1
	Max (Year)	9.07 (1932)	11.31 (2018)	7.39 (2002)	**17.64 (1922)**	16.96 (2015)	11.58 (1928)	11.13 (1973)	10.33 (1915)	12.69 (2018)	15.66 (2018)	9.86 (2015)	8.75 (1991)	62.6 (2015)
	Min (Year)	T (1986)	.01 (1918)	.02 (1925)	.11 (1987)	.22 (1911)	T (1952)	0 (1993)	0 (2000)	.06 (2014)	T (1975)	0 (1903)	T (1950)	0
	24hr Max (Year)	4.27 (2012)	4.72 (2018)	4.39 (1977)	4.55 (1957)	5.34 (1989)	3.98 (2017)	4.01 (2004)	4.05 (1976)	**8.11 (2018)**	5.91 (1959)	4.78 (2015)	4.22 (1991)	8.11 (2018)
Snow (in)	Avg	0.3	0.9	0.2	0	0	0	0	0	0	0	T	0.3	1.7
	Max (Year)	12.1 (1964)	**13.5 (1978)**	6.1 (1942)	T (2007)	0	0	0	0	0	T (1993)	5.0 (1976)	5.5 (1898)	**17.6 (1977–78)**

ACKNOWLEDGMENTS

First and foremost, I want to thank all of the people who so candidly and graciously shared their stories of survival. These were painful memories of violent storms and devastating weather events that changed their lives forever. Their powerful testimonies open a window of understanding on these forces of nature and are a vivid reminder of the need to be vigilant when watches and warnings are issued.

I particularly want to thank Gary Tucker, Pam Russell, Jacqui Bloomquist-Taylor, Donna Hanna, and Gene Yates for sharing their frightening ordeals and stories of the December 26, 2015 tornado outbreak. A heartfelt thanks to Christy Green and Laura Williams for telling their terrifying tales after powerful EF-4 tornadoes leveled their homes and nearly took their lives, and to Blake Wiley and Nancy Fluce, who allowed me into their heavily damaged homes and shared their harrowing encounters of the October 20, 2019 Dallas tornado. I would like to express my gratitude to the England family for allowing me to share the story of their narrow escape from the devastating flash flooding that swamped their home in the middle of the night. Thanks to Gene Shoemake and Mike Evans for documenting their close encounters with the 2000 Fort Worth tornado and the 1957 Dallas tornado, and to Ronnie Ranft for his unique perspective on what it was like to fight the raging Possum Kingdom Complex wildfire. I would also like to thank Richard Laver for his incredible survivor account of the Delta 191 crash that took the life of his father and 134 others. His courageous journey to reclaim his life, and his exhaustive efforts to turn tragedy into triumph by saving the life of his beautiful daughter Kate, are words of inspiration for all of us.

I also had the privilege of interviewing several of the leading scientists and researchers in the field of meteorology who have dedicated a large portion of their lives toward understanding this fascinating and mystical science. I owe my gratitude to them for sharing their knowledge, insight, and expertise. They include Texas State Climatologist John Nielsen-Gammon; Professor Paul Markowski from Penn State University; Josh Wurman from the Center for Severe Weather Research; Tim Marshall of Haag Engineering; Tom Bradshaw, Meteorologist in Charge at the National Weather Service Forecast Office in Fort Worth; and my long-time colleagues Evan Andrews and Clarice Tinsley at FOX 4. I want to thank Vicki and Charles Lester and Chip Mahaney for their help in documenting the events from the 1957 Dallas tornado, and Tonya Hunter for sharing her time and expertise in public safety. I am grateful to have had the opportunity to interview Tim Samaras prior to his untimely passing. His pioneering research enabled us to peer inside the most powerful storm on the planet but cast a spotlight on the dangers of storm chasing.

There are many valuable contributors who graciously allowed me to share their photos, videos, illustrations, and animations to help explain these complex scientific concepts and bring historical weather events to life. These talented people include Brent Isom, Mike Warner, Bryan Snider, Hank Schyma, Mike Mezeul II, Michael Beard, Andy Luten, Martin Lisius, Kevin Brown, Maverick Drone and Photography, Steve Horstmeyer, Steven Butler, and the staff and management at Exposure Labs. (FOX 4 video clips: Footage courtesy of FOX 4 News, NW Communications of Texas, Inc. All Rights Reserved.) Special thanks to Herb Stein, Troy Glover, Hollie Hernandez, Tim Marshall, Jeannine Shaw, Robert Howington, Melanie Brownrigg, Diana Marie, Doug Kovach, Jamesa Larimer, Brad

Rivers, Mark Bishop, Steve Reeves, NOAA, NASA, Texas Tech University, the University of Nebraska at Lincoln, the University of Colorado Boulder, the Center for Collaborative Sensing of the Atmosphere, Climate Central, the National Snow and Ice Data Center, the US Geological Survey, the National Science Foundation, the Center for Severe Weather Research, the University Corporation for Atmospheric Research, and the Federal Aviation Administration for additional photos, videos, and animations. Finally, all of the weather graphics I created for the book were composed using the MAX graphics system from The Weather Company.

I'd be remiss not to mention the many fine colleagues and managers I have had the pleasure of working with over the years. I especially want to thank Eli Jacks, my first boss at the Techniques Development Lab, for his patience and tutelage, and Jeff Hardin, who thankfully saw potential in a very "green" meteorologist and recommended me for my first on-air job at WECT-TV. I am grateful for the opportunity to work with meteorologists Chris Hohmann, Bill Reh, Doug Hill, Evan Andrews, and Ron Jackson, who helped mentor and support me throughout my career. It truly takes a village to provide hours of continuous, live coverage during breaking weather events in a business environment that can best be described as controlled chaos. I am grateful to all the talented meteorologists who have worked beside me in the trenches and to the dedicated professionals who work feverishly behind the scenes to support us and keep us on the air. The volunteers that make up the FOX 4 Storm Force are truly some of the unsung heroes who put themselves in harm's way to get eyes on the storm and provide real-time reports from the field. Finally, the staff at the National Weather Service Forecast Office in Fort Worth play a vital role in helping us get lifesaving information to our viewers during significant weather events. Their commitment and dedication to public safety and serving the great citizens of Texas is truly admirable and we are fortunate to have them on our team.

Paul Spiegelman provided valuable guidance and advice and served as a sounding board throughout the process. My colleagues Evan Andrews and

Alberto Romero also deserve special recognition for their research and graphics assistance and for recalling many of the details from our extended severe-weather coverage.

Many thanks to Ava Coibion for meticulously editing my manuscript and for making excellent comments and suggestions.

A deep debt of gratitude to Stephanie Holan for providing valuable feedback, assistance with editing and bringing photos to life, and serving as the inspiration for the use of QR codes. Thank you for your support and encouraging me to make this book a reality. I also want to thank all the creative and dedicated members of the Scribe Tribe who guided me and supported me through the publishing process. Special thanks to Erica, Erin, Cindy, Matt, Hal, and Josh for fulfilling my countless requests and being so flexible.

Finally, I'd like to thank my beautiful family for putting up with the demands of a cranky author and for their understanding of my many long nights spent at the computer. To Kerry, Logan, Brooke, and Carys, I am eternally grateful for their love and patience during this two-year-long research and writing adventure.

RESOURCES

1. DAY-AFTER-CHRISTMAS TORNADO TRAGEDY

National Weather Service Forecast Office, Fort Worth, "North and Central Texas December 26, 2015 Tornado Outbreak," December 2015, https://www.weather.gov/fwd/dec26tornadoes.

2. WAIT FOR IT

NOAA/National Weather Service JetStream Online Resources, "Longwaves and Shortwaves," https://www.weather.gov/jetstream/longshort.

Iowa State University/Iowa Environmental Mesonet Automated Data Plotter.

Texas Almanac/Environment/Texas Plant Life, https://texasalmanac.com/topics/environment/texas-plant-life.

C. Donald Ahrens, *Meteorology Today: An Introduction to Weather, Climate, and the Environment* (California: Brooks/Cole, 2013), 199–227.

3. WEATHER FORECASTING

Christopher Klein, "The Weather Forecast that Saved D-Day," *History.com*, June 5, 2019, https://www.history.com/news/the-weather-forecast-that-saved-d-day.

Steven Seman, "When Weather Made History," Penn State Department of Meteorology and Atmospheric Science, Meteo 3 Class Notes, https://www.e-education.psu.edu/meteo3/node/2281.

Ralph Jewell, "The Bergen School of Meteorology: The Cradle of Modern Weather Forecasting," *Bulletin of the American Meteorological Society* 62, no. 6, (June 1981): 824–830.

Lou Wicker, "Everything You Need to Know about Numerical Weather Prediction in About 100 Minutes," METR 5004 *Lecture for OU Department of Meteorology*, Fall 2013, http://weather.ou.edu/~scavallo/classes/metr_5004/f2013/lectures/NWP_LecturesFall2013.pdf.

Chris Woodford, "Computer Models," October 9, 2018, https://www.explainthatstuff.com/how-computer-models-work.html.

NOAA/National Weather Service JetStream Online Resources, "Weather Models," https://www.weather.gov/jetstream/models.

Peter Dizikes, "When the Butterfly Effect Took Flight," *MIT Technology Review*, February 22, 2011, https://www.technologyreview.com/s/422809/when-the-butterfly-effect-took-flight/.

4. THUNDERSTORMS, SUPERCELLS, AND TORNADOGENESIS

Anders Clark, "William Rankin: The Man Who Survived Flying through a Thunderstorm," *Disciples of Flight*, February 12, 2015, https://disciplesofflight.com/william-rankins-story/.

William H. Rankin, *The Man Who Rode the Thunder* (New Jersey: Prentice-Hall, 1960).

Royal Meteorological Society, "Luke Howard and Cloud Names," http://archive.rmets.org/weather-and-climate/observing/luke-howard-and-cloud-names.

Robert Davies-Jones, "A Review of Supercell and Tornado Dynamics," *Atmospheric Research* 158 (May 2015): 274–291.

David M. Schultz et al., "Tornadoes in the Central United States and the Clash of Air Masses," *Bulletin of the American Meteorological Society* (November 2014): 1706–1711.

National Weather Service, Louisville, KY, "Supercell Thunderstorm Structure and Evolution," in *Scientific Training Document*, chapter 17. https://www.weather.gov/media/lmk/soo/Supercell_Structure.pdf.

Paul M. Markowski and Yvette P. Richardson, "What We Know and Don't Know about Tornado Formation," *Physics Today* 67, no. 9 (September 2014): 26–31.

Brice E. Coffer and Matthew D. Parker, "Simulated Supercells in Nontornadic and Tornadic Environments," *Monthly Weather Review* 145 (December 2016): 149–180.

Josh Wurman et al., "In Situ, Doppler Radar, and Video Observations of a Tornado and the Wind-Damage Relationship," *Bulletin of the American Meteorological Society* (June 2013): 835–846.

5. TORNADO OUTBREAKS IN NORTH TEXAS

Greg Pierce, "The Worst Storm in Dallas History: The 1957 Dallas Tornado," *Yesterday's News*, University of North Texas Libraries, March 23, 2017, https://blogs.library.unt.edu/yesterdays-news/2017/03/23/the-worst-storm-in-dallas-history-the-1957-dallas-tornado/.

National Weather Service, Fort Worth, TX, "Remembering the Dallas Tornadoes of April 2, 1957," https://www.weather.gov/fwd/dallastor50year.

Rachel Williams, "60 Years Ago This Month, Oak Cliffers Stood Outside to Watch a Deadly F3 Tornado," *Dallas Observer*, April 19, 2017, https://www.dallasobserver.com/arts/the-second-deadliest-tornado-in-dfw-history-struck-oak-cliff-60-years-ago-this-month-9383521.

Chip Mahaney and Tim Marshall, "April 2, 1957: Dallas' Date with Disaster," *Storm Track*, May/June 1997, https://stormtrack.org/library/historic/dal57.htm.

Walter H. Hoecker et al., "The Tornadoes at Dallas, Texas, April 2, 1957," US *Department of Commerce Weather Bureau Research Paper* 41 (1960): 184.

Hank Schyma, "Scariest Tornado Ever," August 10, 2017, video, 10:16, https://www.youtube.com/watch?v=tej3zK91UVQ.

6. EYEING THE STORM

G.K. Grice et al., "The Golden Anniversary Celebration of the First Tornado Forecast," *Bulletin of the American Meteorological Society* 80, no. 7 (July 1999), https://journals.ametsoc.org/doi/pdf/10.1175/1520-0477%281999%29080%3C1341%3ATGACOT%3E2.0.CO%3B2.

Laura Clark, "Why Forecasters Were Once Banned from Using the Word Tornado," *Smithsonian Magazine*, March 25, 2015, https://www.smithsonianmag.com/smart-news/why-forecasters-were-once-banned-using-word-tornado-180954742/.

Roger C. Whiton and Paul L. Smith, "History of Operational Use of Weather Radar by US Weather Services. Part 1: The Pre-NEXRAD Era," *Weather and Forecasting* 13 (June 1, 1998), https://journals.ametsoc.org/doi/full/10.1175/1520-0434%281998%29013%3C0219%3AHOOUOW%3E2.0.CO%3B2.

Charles Doswell III et al., "Storm Spotting and Public Awareness Since the First Tornado Forecasts of 1948," *Weather and Forecasting* 14 (August 1999): 544–557, https://www.nssl.noaa.gov/users/brooks/public_html/papers/doswellmollerbrooks.pdf.

Joey Picca, "Dual-Pol Radar: Tornado Debris Signature," NOAA *Severe Thunderstorm Forecasting Lecture Series*, May 15, 2017, video, https://www.youtube.com/watch?v=RKUP46UE-1I&feature=youtu.be.

Haonan Chen et al., "Principles of High Resolution Radar Network for Hazard Mitigation and Disaster Management in an Urban Environment," *Journal of the Meteorological Society of Japan* 96A (January 2018): 119–139.

Josh Wurman et al, "The Role of Multiple-Vortex Tornado Structure in Causing Storm Researcher Fatalities," *Bulletin of the American Meteorological Society* 95, no. 1 (January 2014): 31–45, https://journals.ametsoc.org/doi/full/10.1175/BAMS-D-13-00221.1.

NOAA/National Severe Storms Laboratory, "FACETs: Forecasting a Continuum of Environmental Threats," https://www.nssl.noaa.gov/projects/facets/.

Kevin Skow, "NWS WSR-88D Radar Fundamentals," Class Lecture at Iowa State University, 2013, https://www.meteor.iastate.edu/classes/mt432/lectures/ISURadarTalk_NWS_2013.pdf.

NOAA/National Severe Storms Laboratory, "Warn-on-Forecast," Fact Sheet, 2015, https://www.nssl.noaa.gov/news/factsheets/WoF_2015.pdf.

7. MEGA HAIL AND DOWNBURST WINDS

NOAA/National Weather Service, "Thunderstorm Hazards—Hail," *JetStream Online Resources*, https://www.weather.gov/jetstream/hail.

Eli J. Dennis and Matthew R. Kumjian, "The Impact of Vertical Wind Shear on the Growth of Hail in Simulated Supercells," *Journal of the Atmospheric Sciences* 74, no. 3 (March 2017): 641–663.

Insurance Council of Texas, "Costliest Texas Storms: Insured Losses—Actual Dollars," https://www.insurancecouncil.org/costliest-texas-storms-insured-losses/.

National Weather Service, Fort Worth, TX, "Supercell Thunderstorms Create Swath of Damage and Injury April 5, 2003," April 2003, https://www.weather.gov/fwd/april5_2003_supercells.

NASA Global Hydrology Resource Center, "A Lightning Primer," https://ghrc.nsstc.nasa.gov/home/lightning/home/primer/primer2.html.

National Severe Storms Laboratory Learning Resources, "Severe Weather 101: Lightning," https://www.nssl.noaa.gov/education/svrwx101/lightning/faq/.

Federal Aviation Administration—Lessons Learned from Civil Aviation Accidents, "Delta Flight 101 Accident Overview," https://lessonslearned.faa.gov/ll_main.cfm?TabID=2&LLID=32&LLTypeID=2.

"2-August 1985-Delta 191," *Tailstrike.com, Cockpit Voice Recorder Database*, https://www.tailstrike.com/020885.htm.

NOAA/Storm Prediction Center, "The Texas Derecho of 1989," https://www.spc.noaa.gov/misc/AbtDerechos/casepages/may41989page.htm.

Timothy P. Marshall et al., "Hail Damage Threshold Sizes for Common Roofing Materials," *21st Conference on Severe Local Storms*, San Antonio, TX, August 2002, https://www.researchgate.net/publication/327022658_HAIL_DAMAGE_THRESHOLD_SIZES_FOR_COMMON_ROOFING_MATERIALS.

Jim D. Koontz, "Effects of Hail on Residential Roofing Products," *Third International Symposium on Roofing Technology*, Montreal, Canada, April 1991.

8. FLASH FLOODING, HURRICANES, AND TROPICAL DELUGES

US Army Corps of Engineers—Fort Worth District, "Dallas Floodway Timeline," https://www.swf.usace.army.mil/Portals/47/docs/PAO/DF/PDF/Dallas_Floodway_ Timeline_1908-2013.pdf.

Hometown by Handlebar (blog), May 17, 2019, https://hometownbyhandlebar.com/ ?p=6771.

Bud Kennedy, "Fort Worth's Flood," *Fort Worth Star Telegram*, August 28, 2017.

John N. Furlong et al., "History of the Dallas Floodway," *American Society of Civil Engineers Paper*, Fall 2003, http://citeseerx.ist.psu.edu/viewdoc/ download?doi=10.1.1.729.4470&rep=rep1&type=pdf.

University of Rhode Island, "Hurricanes: Science and Society," 2010–2015, http:// hurricanescience.org/science/science/hurricanestructure/.

Kevin E. Trenberth et al., "Hurricane Harvey Links to Ocean Heat Content and Climate Change Adaptation," *Earth's Future* 6, no. 5 (May 2018): 730–744, https:// agupubs.onlinelibrary.wiley.com/doi/full/10.1029/2018EF-000825.

James P. Kossin, "A Global Slowdown of Tropical Cyclone Translation Speed," *Nature International Journal of Science* 558 (June 6, 2018): 104–107, https://www. researchgate.net/publication/325605246_A_global_slowdown_of_tropical- cyclone_translation_speed.

C. Donald Ahrens, Meteorology Today: An Introduction to Weather, Climate, and the Environment (California: Brooks/Cole, 2013), 424–455.

9. HEAT WAVES, DROUGHTS, AND THE WATER SUPPLY

Jerome H. Greenberg et al., "The Epidemiology of Heat-Related Deaths, Texas 1950, 1970–1979, and 1980," *American Journal of Public Health* 73, no. 7 (July 1983), https://ajph.aphapublications.org/doi/pdf/10.2105/AJPH.73.7.805.

Justice Jones et al., "2011 Texas Wildfires: Common Denominators of Home Destruction," Texas A&M Forest Service (2011), 1–52, https://tfsweb.tamu.edu/ uploadedFiles/TFSMain/Preparing_for_Wildfires/Prepare_Your_Home_for_ Wildfires/Contact_Us/2011%20Texas%20Wildfires.pdf.

Cindy Devone-Pacheco, "2011 Wildfires: Two Perspectives," *FireRescue* 12, no. 6 (December 1, 2011), https://firerescuemagazine.firefighternation.com/2011/12/01/ 2011-texas-wildfires-two-perspectives/.

Water Data for Texas, TWDB Reports, https://waterdatafortexas.org/drought/twdb- reports.

Texas Water Development Board, "2016 Region C Water Plan Volume 1 Main Report," http://www.twdb.texas.gov/waterplanning/rwp/plans/2016/c/Region_C_ 2016_RWPV1.pdf?d=4801.129999803379.

Norman Johns, "Evaporation: A Loss for Humans and Wildlife in Texas," *Texas Living Waters Project Blog*, September 24, 2014, https://texaslivingwaters.org/evaporation-loss-humans-wildlife-texas/.

Sharlene Leurig, "Building a Water-Resilient Future for All Texans," *Austin American-Statesman*, September 16, 2019, https://www.statesman.com/opinion/20190916/building-water-resilient-future-for-all-texans.

10. WINTER WEATHER

Josie Garthwaite, "The Polar Vortex: How Does it Work?" *Futurity*, February 6, 2019, https://www.futurity.org/polar-vortex-1975412/.

National Weather Service, Fort Worth, Texas, "North and Central Texas Snowfall Events for Winter 2009-2010," https://www.weather.gov/fwd/wintersnow09-10.

Jeff Sullivan, "STAR: Frozen in Time—Thanksgiving 1993 Memories," *DallasCowboys.com*, November 27, 2013, https://www.dallascowboys.com/news/star-frozen-in-time-thanksgiving-1993-memories-344656.

Mike Berardino, "Chicken Soup, Bitter Cold, and a Notre Dame Miracle at the 1979 Cotton Bowl," *Indianapolis Star*, December 13, 2018, https://www.indystar.com/story/sports/college/notre-dame/2018/12/13/how-notre-dame-clipped-houston-35-34-1979-cotton-bowl/2278002002/.

11. BE PREPARED!

NOAA/National Weather Service Central Region Headquarters, "NWS Central Region Service Assessment—Joplin, Missouri Tornado—May 22, 2011," July 2011, https://www.weather.gov/media/publications/assessments/Joplin_tornado.pdf.

Federal Emergency Management Agency, "Taking Shelter from the Storm—Building a Safe Room for Your Home or Small Business," *FEMA P-320*, 4th edition, December 2014, https://www.fema.gov/media-library-data/1418837471752-920f09bb8187ee15436712a3e82ce709/FEMA_P-320_2014_508.pdf.

Lynn Walker, "Memories of Deadly Terrible Tuesday Remain 40 Years After Tornado's Wrath," *Times Record News*, April 10, 2019, https://www.timesrecordnews.com/story/news/local/2019/04/10/terrible-tuesday-wichita-falls-tornado-1979-memories/3402945002/.

Department of Homeland Security, "Build a Kit—Basic Disaster Supplies Kit," https://www.ready.gov/kit.

Benchmark H. Harris, "Tornado Shelters in Schools," *Structure Magazine*, September 2016, https://www.structuremag.org/?p=10396.

12. WHAT'S NEXT?

NASA Earth Observatory, "World of Change: Global Temperatures," https://earthobservatory.nasa.gov/world-of-change/DecadalTemp.

Intergovernmental Panel on Climate Change, "Special Report: Global Warming of 1.5°C," https://www.ipcc.ch/sr15/.

Rebecca Lindsey, "Climate and Earth's Energy Budget," *NASA Earth Observatory*, January 14, 2009, https://earthobservatory.nasa.gov/features/EnergyBalance.

NOAA National Centers for Environmental Information, "What Are Proxy Data?" https://www.ncdc.noaa.gov/news/what-are-proxy-data.

National Science Foundation Ice Core Facility, "About Ice Cores," https://icecores.org/about-ice-cores.

Michon Scott, "Antarctica is Colder than the Arctic, But It's Still Losing Ice," *NOAA Climate.gov*, March 12, 2019, https://www.climate.gov/news-features/features/antarctica-colder-arctic-it%E2%80%99s-still-losing-ice.

American Chemical Society, ACS Climate Science Toolkit, "It's Water Vapor, Not the CO_2," https://www.acs.org/content/acs/en/climatescience/climatesciencenarratives/its-water-vapor-not-the-co2.html.

Marshall Shepherd, "Water Vapor vs. Carbon Dioxide: Which Wins in Climate Warming?" *Forbes Magazine*, June 20, 2016, https://www.forbes.com/sites/marshallshepherd/2016/06/20/water-vapor-vs-carbon-dioxide-which-wins-in-climate-warming/#8c2348e3238f.

Andrew Dessler, "What We Know about Climate Change," *Testimony before the US Senate Committee on Environment and Public Works*, January 16, 2014, https://www.epw.senate.gov/public/_cache/files/2/6/26edecac-2c6f-4f8e-ab90-962a7d074d06/01AFD79733D77F24A71FEF9DAFCCB056.11614hearingwitnesstestimonydessler.pdf.

"Climate Forcing," *NOAA Climate.gov*, https://www.climate.gov/maps-data/primer/climate-forcing.

Mathew J. Owens et al., "The Maunder Minimum and the Little Ice Age: an Update from Recent Reconstructions and Climate Simulations," *J. Space Weather Space Clim* 7 (December 4, 2017), https://www.swsc-journal.org/articles/swsc/full_html/2017/01/swsc170014/swsc170014.html.

Kevin Krajick, "In Ancient Rocks, Scientists See a Climate Cycle Working Across Deep Time," *State of the Planet*, Earth Institute/Columbia University, May 7, 2018, https://blogs.ei.columbia.edu/2018/05/07/milankovitch-cycles-deep-time/.

Jason Wolfe, "Volcanoes and Climate Change," *NASA EarthData Sensing Our Planet*, May 15, 2019, https://earthdata.nasa.gov/learn/sensing-our-planet/volcanoes-and-climate-change.

Renee Cho, "How We Know Today's Climate Change Is Not Natural," *State of the Planet*, Earth Institute/Columbia University, April 4, 2017, https://blogs.ei.columbia.edu/2017/04/04/how-we-know-climate-change-is-not-natural/.

David Furphy, "What on earth is an RCP?" *Medium,* September 29, 2013, https://medium.com/@davidfurphy/what-on-earth-is-an-rcp-bbb206ddee26.

Geert Jan van Oldenborgh et al., "Attribution of Extreme Rainfall from Hurricane Harvey, August 2017," *Environmental Research Letters* 12, no. 12 (December 13, 2017), https://iopscience.iop.org/article/10.1088/1748-9326/aa9EF-2/meta.

David Reidmiller et al., "Fourth National Climate Assessment: Volume 11: Impacts, Risks, and Adaptations in the United States," US *Global Change Research Program,* Washington, DC, 2018, https://nca2018.globalchange.gov/ .

Aylin Woodward, "11 Ways We Could Geoengineer the Planet to Reverse Climate Change," *Business Insider,* April 20, 2019, https://www.businessinsider.com/geoengineering-how-to-reverse-climate-change-2019-4.

INDEX

ABOUT THE AUTHOR

Dan Henry is the Chief Meteorologist at FOX 4 in Dallas, Texas. A five-time Emmy award winner, Dan has covered the most notable weather events of the past several decades, from the East Coast Blizzard of 1996 to the deadly tornado outbreak in the DFW Metroplex on December 26, 2015. Dan has earned the prestigious Certified Broadcast Meteorologist designation from the American Meteorological Society. He regularly speaks to schools, businesses, churches, and civic organizations on severe weather preparedness. To connect, check out DanHenryWeather.com.

CPSIA information can be obtained
at www.ICGtesting.com
Printed in the USA
BVHW061420120620
581327BV00002B/2